Oyster Culture

Oyster Culture

George C. Matthiessen

Connecticut, USA

Fishing News Books

Blackwell
Science

A division of Blackwell Science Ltd
Editorial Offices:
Osney Mead, Oxford OX2 0EL
25 John Street, London WC1N 2BS
23 Ainslie Place, Edinburgh EH3 6AJ
350 Main Street, Malden
MA 02148 5018, USA
54 University Street, Carlton
Victoria 3053, Australia
10, rue Casimir Delavigne
75006 Paris, France

Other Editorial Offices:

Blackwell Wissenschafts-Verlag GmbH
Kurfürstendamm 57
10707 Berlin, Germany

Blackwell Science KK
MG Kodenmacho Building
7-10 Kodenmacho Nihombashi
Chuo-ku, Tokyo 104, Japan

Iowa State University Press
A Blackwell Science Company
2121 S. State Avenue
Ames, Iowa 50014-8300, USA

First published 2001

Set by Best-set Typesetter Ltd., Hong Kong

DISTRIBUTORS

Marston Book Services Ltd
PO Box 269
Abingdon
Oxon OX14 4YN
(*Orders:* Tel: 01865 206206
Fax: 01865 721205
Telex: 83355 MEDBOK G)

USA and Canada
Iowa State University Press
A Blackwell Science Company
2121 S. State Avenue
Ames, Iowa 50014-8300
(*Orders:* Tel: 800-862-6657
Fax: 515-292-3348
Web www.isupress.com
email: orders@isupress.com)

Australia
Blackwell Science Pty Ltd
54 University Street
Carlton, Victoria 3053
(*Orders:* Tel: 03 9347 0300
Fax: 03 9347 5001)

A catalogue record for this title is available from the British Library

ISBN 0-85238-279-0

Library of Congress
Cataloging-in-Publication Data is available

For further information on Fishing News Books, visit our website:
http://www.blacksci.co.uk/fnb/

Contents

Preface

A few years ago, I happened to be in New York's Grand Central Station and, to kill some time before catching a train, wandered into the well-patronized Oyster Bar. Glancing at the menu, I was amazed by the price of oysters on the half-shell, which, at that time, exceeded $2.00 apiece. Nevertheless, it was gratifying to note that, of the ten or so different oysters being featured on the menu and identified by name, three – Cotuit, Cuttyhunk and Fishers Island – had originated from our small oyster farm. Oysters we had nursed from embryo through adolescence were making it to the big time.

My interest in shellfish farming began during college when I worked one summer at the Woods Hole Oceanographic Institution under the late Harry Turner, a well-known shellfish biologist. He and Dr Alfred Redfield, the assistant director of the Institution at that time, were attempting to farm soft clams (*Mya arenaria)* in small intertidal plots on the north shore of Cape Cod. This interest was further stimulated at graduate school by the book *Living Resources of the Sea*, by Dr Lionel A. Walford, who wrote the following:

> 'There are long stretches of coast about the world where land and sea are not so sharply distinct as they are at rocky shore; where, instead, the two merge gradually in an irregular and sometimes intricate edging of estuaries, sloughs, lagoons and mud flats, brackish swamps, and fringing islands. This transitional area where the land reaches out to the sea and the sea into the land is one of the most interesting of all marine ecosystems, and perhaps from a fishery viewpoint potentially the most valuable. In some places it is richer than the richest farm country . . .'[P.1].

The salt ponds that dot our coastlines are interesting examples of the brackish-water environment to which Walford referred, and their uniqueness has been exploited by fish and shellfish culturists for centuries. Such ponds may offer protection from excessive wave action and strong tidal currents, and the pond environment can often be controlled to a degree by use of tide gates or other barriers that control the inflow of ocean water. Predators may be reduced or excluded in this manner, and harvests have in some cases been increased by supplementary enrichment with fertilizer, a strategy that would be impractical in the open ocean. A further advantage to pond culture is the fact that the planktonic larvae of valuable species established in the pond are apt to be retained rather than swept by tidal currents to inaccessible areas offshore. The concept of systematically managing a small piece of the ocean, such as a salt pond, for food production was for me an exciting one.

Pond culture of marine animals dates back many centuries. During the second century BC, oysters were being grown in Lake Lucrinus, near Naples, by Roman oyster culturists[P.2]. Other examples of ancient brackish water pond culture include the mullet farms of the Romans, Egyptians and Hawaiians[P.3] and the milkfish farms in South-east Asia[P.4], which date back at least to AD 1400 in Indonesia. Today, brackish water shrimp and finfish culture continues to be practiced in many countries, including Italy, China, Taiwan, Hawaii, Israel, Phillipines, Indonesia and parts of South America.

This book is about oyster culture, as we have practiced it in a brackish-water pond for nearly forty years and as it is practiced in different parts of the world. Descriptions of culture methods other than ours are based upon personal interviews, site visits, and literature reviews. Countries that have been visited for this purpose include Japan, China, Australia, New Zealand, France, Great Britain and Mexico.

Acknowledgements

During the course of gathering information for this book, I have been helped by a good many people. Among those in foreign countries, the following individuals have been of particular help: Harry Shaw, Baugham Wisely, Ian Cameron, Neville Phillips and Trevor Dix (Australia); Les Curtin (New Zealand); Osamu Fukuhara and Tomohiro Kimura (Japan); Qiu Li-Qiang (People's Republic of China); Maurice Heral, Edouard His and Yves LeBorgne (France); Mark Dravers (Guernsey); Clive Askew and John Bayes (Great Britain); Eric Baqueiro and Rene Islas Olivares (Mexico).

There are also a good many individuals involved in shellfish culture and/or research with whom I have had the pleasure of working and/or who have provided valuable help to Ocean Pond Corporation one way or another over the years. In this regard, I would particularly like to thank Richard Toner, Richard Nelson, David Relyea, Hal Haskin, Susan Ford and, among Ocean Pond Corporation's ex-employees, Tim Putnam, Charles LeCour and Paul Evans, who spent more than their share of time working at the pond. I am especially grateful to Jonathan Davis and Steve Malinowski, whose help and suggestions have been invaluable.

A great deal of thanks goes to my father, Erard A. Matthiessen, who was the founder of Ocean Pond Corporation and who managed the operation so successfully during its early years; to my brother, Peter Matthiessen, for his helpful suggestions on this book; and to my wife, Cis, for the many days of cold, wet and often tedious work that she put into culturing oysters.

Introduction

The process of producing oysters takes various forms. There is a 'wild' fishery, in which the fisherman simply goes out on the water in search of natural, uncultivated populations and harvests whatever he can by rake, tongs, dredge or hand. His input, as far as the oyster's welfare is concerned, is minimal. Then there is a sort of semi-culture whereby the grower assists nature to some degree. This might involve spreading shell over the bottom or suspending strings of shell from rafts or racks in order to obtain natural 'sets' of wild oysters, or controlling natural predators as best he can, or moving the oysters from less favorable growing areas to more favorable ones as the oysters mature. The great majority of oysters produced in the world today are provided in this way. Finally, there is a form of intensive culture, or husbandry, whereby the culturist is involved in all stages of the oyster's conception and growth: the parent stock is carefully selected, the eggs are fertilized and the larvae reared under controlled conditions, and the juveniles are grown to maturity in nets, on strings, in plastic mesh bags, or by whatever means that will insure a high rate of growth and survival. This process is analogous to contemporary agriculture, although not nearly as sophisticated.

Island Pond is a brackish-water pond in southern New England. Oysters have been cultured in, and harvested from, this pond in commercial quantities since 1962, and, in the process, information has been gathered on the biota, hydrography and chemistry of the pond itself as well as on the ecology of the native oysters. When this project first got under way, attempts to practice underwater farming of shellfish had already been considered or initiated in certain coastal ponds in the New England area[I.1,I.2], but the science of marine aquaculture was still in its infancy. Methods of oyster production, for example, differed very little from those of the previous century, and oyster farmers relied heavily upon favorable natural events to ensure a successful crop.

The history of oyster production in Island Pond is a small example of the progress of marine aquaculture from the early 1960s, when 'mariculture' was a relatively new and exciting concept, to the present. Despite the growth of the aquaculture industry in recent decades, the science of marine farming nonetheless remains an uncertain one in many respects. Most aquaculturists experience severe disappointments along the way, encountering obstacles that are often major, sometimes terminal. Perhaps one of aquaculture's attractions is that there are always surprises, often interesting, frequently unpleasant, and that there still remains a great deal of room for technological improvement.

In the early 1960s, oyster production in New England had reached nearly an all-time low. The natural populations once common along the Connecticut shoreline and in parts of Rhode Island, Massachusetts and even Maine, had been either over-harvested or destroyed by hurricanes. In Long Island Sound, once-productive beds had become depleted, and private oyster growers were desperate for seed oysters.

It was at this time that Ocean Pond Corporation was formed for the purpose of utilizing Island Pond on Fishers Island, New York, as a source of seed oysters for the regional industry. I was working on the Island of Martha's Vineyard, doing research on methods of improving the island's shellfishing industry, and went down to Fishers Island in the fall of 1961 to visit my parents. My father, recently retired from architecture but very active in conservation groups, took me to Island Pond, where he had recently acquired some property, to help haul his aluminum rowboat out of the water. When we flipped the boat over, I was amazed by the concentration of oyster spat that covered the bottom. A brief walk around the margins of the pond revealed an extremely dense oyster population, consisting of several different year-classes, and it was obvious that the heavy set that had occurred in 1961 was not an exceptional phenomenon.

According to local lore, an island resident had obtained a bushel of oysters some years ago from an oysterman in Connecticut, had put the oysters in the pond for future use, and had then forgotten all about them. These oysters had evidently reproduced and, in the absence of heavy competition and predation, their progeny had soon established a dense population on the submerged rocks in the pond. Given the current deplorable condition of the oyster industry in New England, a pond such as this seemed to offer a unique opportunity not only for experimental culture but also for commercial production, and within a few months my father had formed the Ocean Pond Corporation, a lease for the shellfishing rights in the pond had been obtained, and a permit for culturing oysters had been granted from the State Department of Environmental Conservation.

Since Ocean Pond Corporation's inception in 1962, many millions of seed oysters have been harvested from the pond and supplied to private growers, some as far away as the Hawaiian Islands. One of the purposes of this book is to provide an account of the company's culture methods that evolved over the years and acquaint prospective oyster growers with some of the frustrations inherent in marine aquaculture. Several Ocean Pond employees over the years – Steve Malinowski, Jonathan Davis, Seth Garfield and Robert Rheault – have subsequently established their own oyster farms. No doubt their experiences working for Ocean Pond have helped them avoid some obstacles but probably have led them to discover new ones as well.

1 Salt Ponds

Coastal bodies of water partially or periodically separated from the ocean by a barrier beach are generally referred to as salt ponds or lagoons. Some ponds may have a permanent opening to the sea and undergo a tidal rise and fall accompanied by a relatively high and constant salt concentration. In others, communication with the sea may be intermittent, as when the pond opening is periodically filled with sand from natural causes and then reopened by wave action. Because of restricted tidal flow, the water within ponds such as these tends to be brackish; the salt concentration may vary appreciably from month to month depending upon local rainfall or periodic influx of sea water as a result of coastal storms or extreme tides.

Island Pond, with restricted access to the sea, falls in the second category. It is located on Fishers Island, New York, a narrow strip of land about 11 kilometers (km) (7 miles) long lying 5 km off the coast of Connecticut in southern New England (Fig. 1.1). Until the 1938 hurricane, Island Pond was a freshwater pond used by Island residents as a supplementary reservoir, sealed off from salt water intrusion by a barrier beach. Wave action and an extreme high tide during the hurricane destroyed the dike that separated Island Pond from Beach Pond, an adjacent and smaller body of water that periodically drained into Block Island Sound via a narrow channel (Fig. 1.2). In the process, large volumes of sea water flooded both ponds. Following repair of the dike, efforts to restore freshwater conditions in Island Pond by pumping out the sea water were unsuccessful, the dike was eventually breached by another storm, and the pond has remained essentially brackish since that time.

Island Pond is approximately 14 hectares (ha) (35 acres) in total surface area, with a maximum depth of about 7 meters (m) (22 ft) when the pond level is low (Fig. 1.3). A peninsula, roughly 3 ha in area, extends from its north shore in a south-westerly direction, nearly creating two separate ponds. In the shoal areas, out to a depth of about 2 m, the bottom sediment is relatively firm and consists of a mixture of sand and silt. With increasing depth, the bottom becomes increasingly soft, grading eventually into a fine black organic ooze in the deepest areas. The pond is relatively sheltered from wind from the north and west by high rocky and wooded bluffs rising from the shore line. The east side of the pond is bordered by an extensive flat meadow, while the south shore line separating the pond from Block Island Sound consists of a sand and rock cobble beach approximately

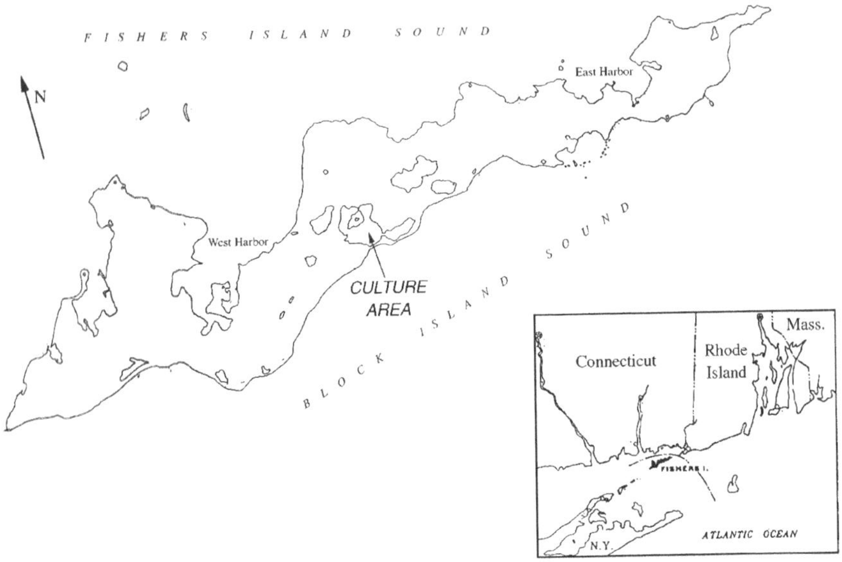

Fig. 1.1 Map of Fishers Island showing location of culture area (Island Pond).

Fig. 1.2 Aerial photograph of Island Pond, Beach Pond, and adjacent freshwater ponds.

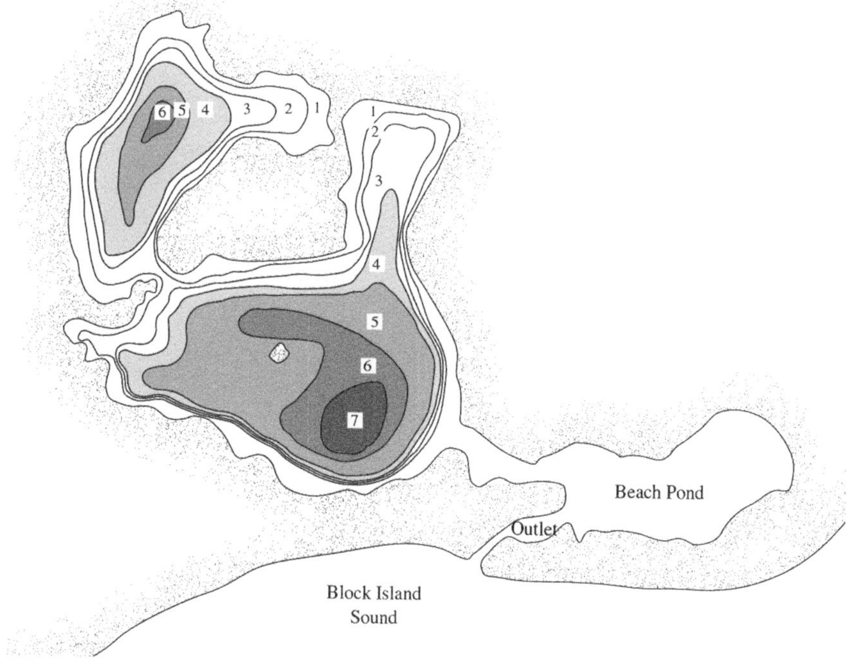

Fig. 1.3 Bottom contours of Island Pond (depth in m).

60 m in width, giving the pond little protection from winds from the south and east (Fig. 1.4). The characteristics of Island Pond, relatively deep, located at the edge of the glacial moraine and surrounded by glacial till for the most part, liken it to a kettle pond depression.

Periodically, during hurricanes or severe winter storms accompanied by extreme high tides, large volumes of sea water enter Island Pond by way of Beach Pond. Under most conditions, however, the outlet between Beach Pond and Block Island Sound either has been closed or was so shallow that very little sea water penetrated into Island Pond even during normal tides. When the pond remained closed for extended periods, the pond level would gradually rise 0.5 m or more above mean tide level as a result of rainfall and land runoff. Recently a more stable and permanent outlet was excavated across the cobble beach between Beach Pond and the Sound. This has not increased the salinity in Island Pond appreciably but has served primarily as a drain that moderates fluctuations in salt concentration. Other than during spring tides, the tidal rise and fall in Island Pond is less than 10 centimeters (cm).

The pond's surface salinity, or salt concentration, normally falls between 16‰ (parts per thousand) and 22‰: on the average, it is about

Fig. 1.4 (a) Aerial photograph of Island Pond (Block Island Sound in background) (photo by John Ski).

Fig. 1.4 (b) Aerial photograph of Island Pond from the south-east (Fishers Island Sound in background) (photo by John Ski).

two-thirds the concentration in Block Island Sound (30–31‰). Immediately after a major intrusion of sea water, however, surface salinities may measure 25‰ or slightly more for several days. Prior to construction of the new opening, the water level in the pond would rise after periods of heavy rain, and the surface salinity would drop to 15‰ or below. Since normal development and survival of oyster (*Crassostrea virginica*) eggs and larvae have been found to decline at salinities below 15‰[1.1], it was customary prior to the spawning season to open the pond to the Sound whenever the salinity had dropped to this level. This allowed the low salinity surface water to drain out and high salinity Sound water to enter during extreme tides.

Opening the pond normally consisted of digging a narrow trench across the barrier beach by hand shovel or bulldozer. With a sufficient head of water in the pond, once the seaward flow began, the pressure of the current would soon excavate a channel up to 6 m or more in width. Within a few days to a week, depending upon prevailing winds, tides and size of the opening, the pond level would drop nearly to mean sea level. Now, with a small but permanent channel, this procedure is no longer necessary.

An interesting feature of Island Pond is the degree to which the pond stratifies during the spring and summer months. Stratification is most pronounced in the south-east sector, where depth is greatest. Here, bottom water temperature may be 10°C colder than at the surface (Fig. 1.5), and the bottom salinity may be 10‰ higher (Fig. 1.6). The summer thermocline and halocline break down rapidly during the fall, the pond becoming essentially isothermal and isohaline during the winter, and stratification does not recur until late spring. During the winter, ice may form over the pond, sometimes to a thickness of more than 20 cm in severe winters, but certain parts of the pond remain open or nearly so, indicating the presence of underwater springs beneath.

As the thermocline develops during the spring, the chemistry of the pond water near the bottom changes as well, with a marked reduction in dissolved oxygen and pH (Figs. 1.7, 1.8), formation of hydrogen sulfide (H_2S) and an increase in the concentrations of certain inorganic nutrients, in particular phosphate (PO_4), silicate (SiO_3) and ammonia (NH_3). How this source of nutrient-rich water is utilized for the purpose of enhancing the growth of algal food for oysters will be described in a later chapter.

The north and west shore lines of the pond, as well as the peninsula, are quite heavily wooded. The dominant species nearest the water is red maple. On higher ground are stands of American beech and red oak, as well as shadbush and staghorn sumac. Much of the undergrowth on the peninsula consists of arrowwood, bayberry, highbush blueberry, honeysuckle, poison ivy, shining sumac, pepperbush and winterberry. The large trees, particularly the maple and oak, contribute large amounts of organic matter to the pond in the form of leaves, which settle to the bottom and decompose.

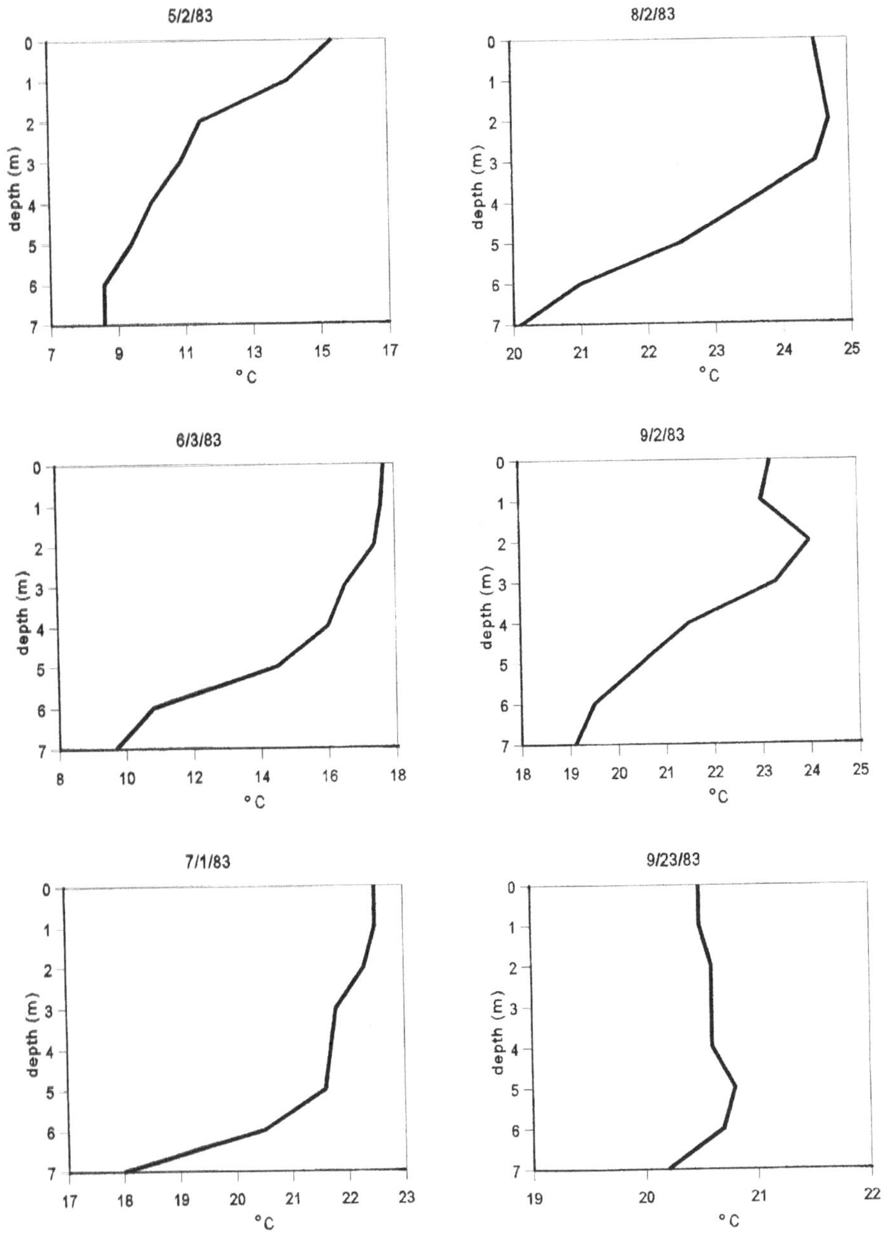

Fig. 1.5 Temperature profiles, Station S-1, Island Pond.

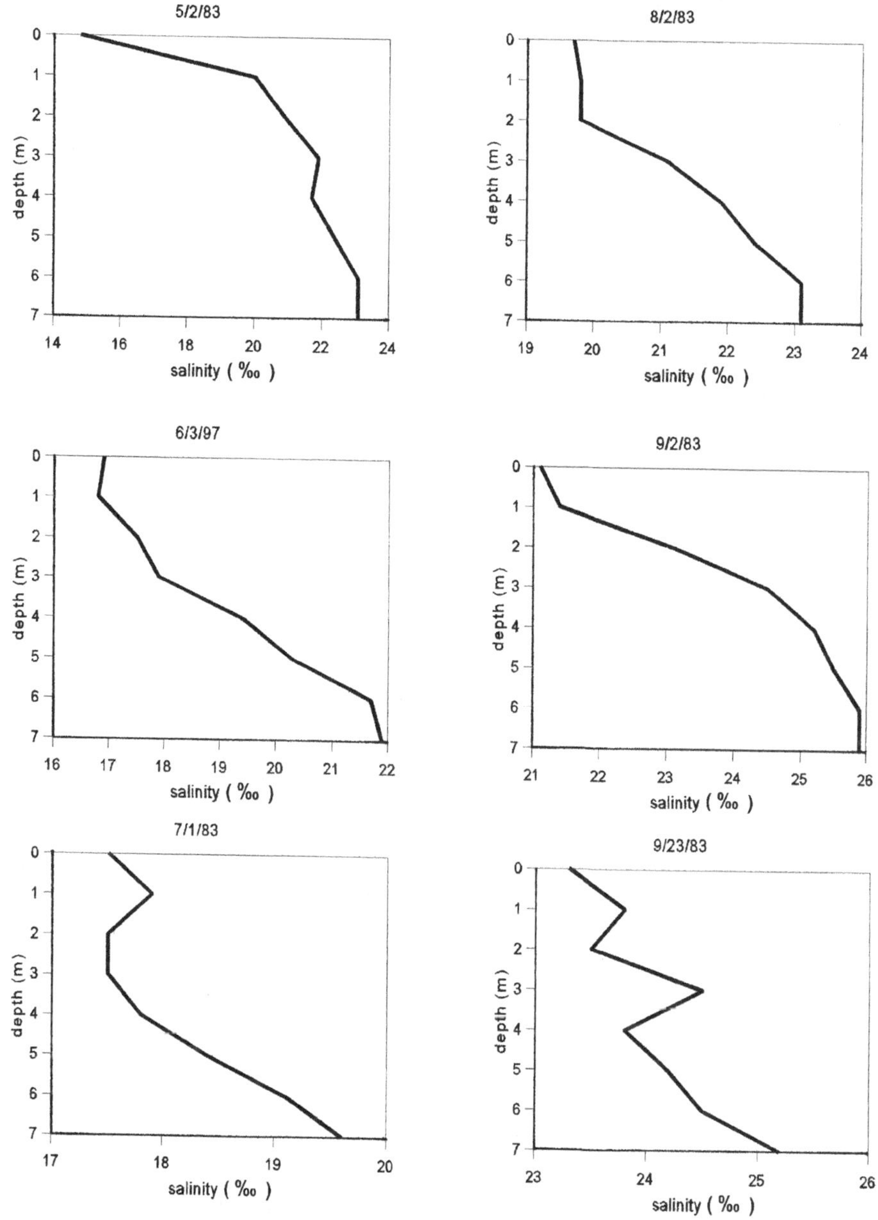

Fig. 1.6 Salinity profiles, Station S-1, Island Pond.

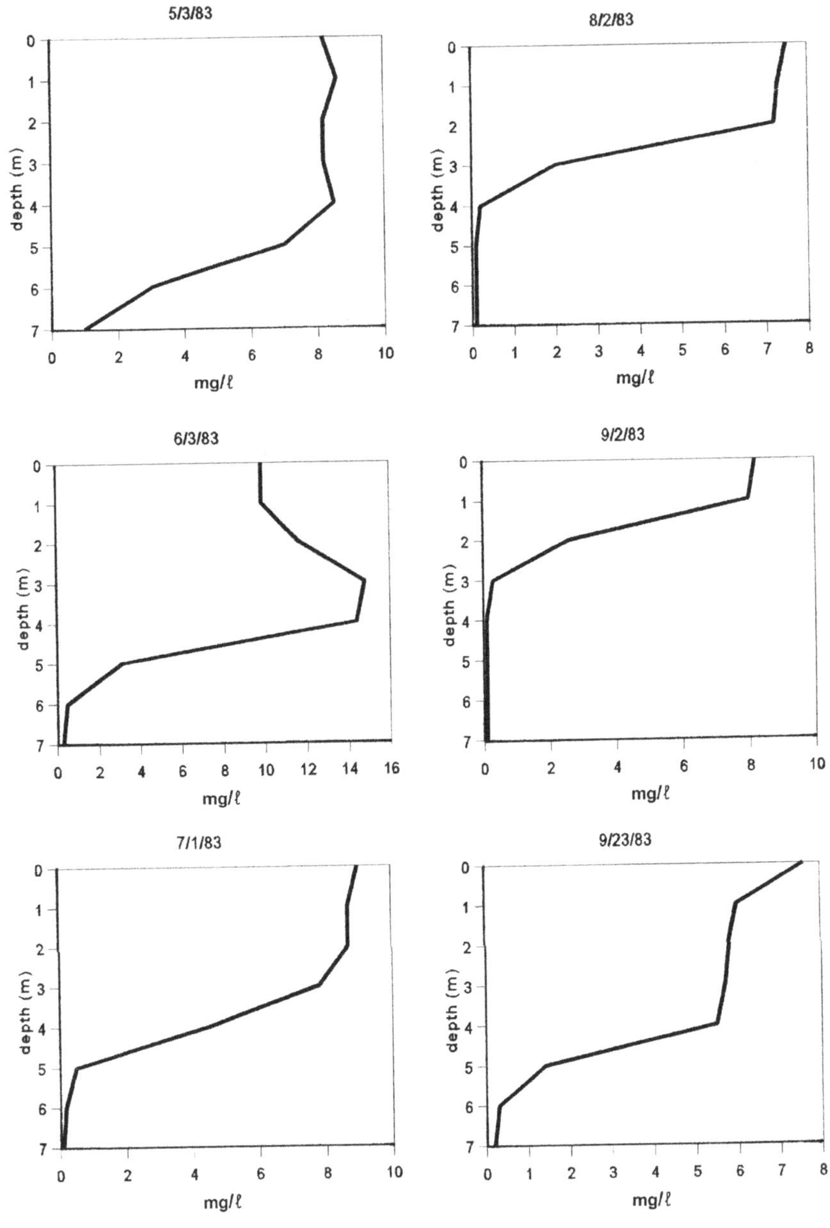

Fig. 1.7 Dissolved oxygen profiles, Station S-1, Island Pond.

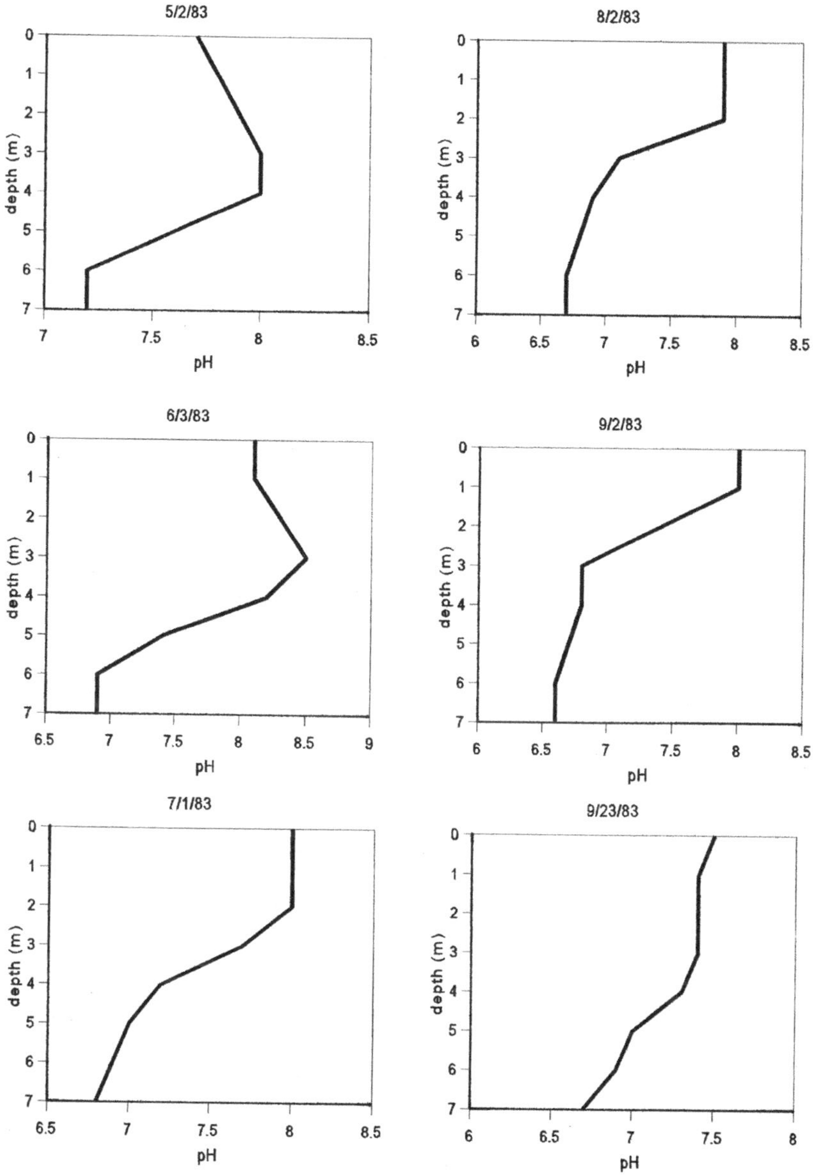

Fig. 1.8 pH profiles, Station S-1, Island Pond.

Aquatic vegetation in the pond itself is limited to shallow areas, i.e. less than 2 m in depth, where the sediment is relatively firm. The dominant species are two spermatophytes, eelgrass (*Zostera marina*), widgeon grass (*Ruppia maritima*), and macroalgae, primarily sea lettuce (*Ulva lactua*). During major storms, the large influx of sea water often brings fragments of other species of marine macroalgae, such as rockweed (*Fucus vesiculosus*) and knotted wrack (*Ascophyllum nodosum*), into the pond, but these oceanic species do not become established, probably because of the pond's relatively low salinities and high water temperatures during summer.

A variety of aquatic birds frequent the pond on a seasonal basis. Winter visitors include the bufflehead, great cormorant, pied-billed grebe and red-breasted merganser. Among the common summer species are the American bittern, double-crested cormorant, American black duck, great egret, snowy egret, gadwall, Canada goose, black-crowned night heron, great blue heron, green heron, mallard and mute swan. Both the black-backed gull and herring gull frequent the pond year-around and nest on the barrier beach. Four other seasonal visitors that may also choose the barrier beach for breeding purposes are the least tern, piping plover, American oystercatcher, and killdeer. Ospreys and kingfishers visit the pond periodically in search of fish, and a family of ospreys has occupied a nest near the edge of the pond for many years.

Populations of finfish and invertebrate organisms in the pond are relatively low in species diversity. This is usually true for brackish water environments, where marked variations in salinity and extremes in temperature tend to exclude many species commonly found in adjacent and more stable oceanic habitats. Typically, a few euryhaline species account for most of the biomass, which can be quite large. The most abundant finfish in Island Pond are the common mummichog (*Fundulus heteroclitus*), Atlantic silverside (*Menidia menidia*), and three-spined stickleback (*Gasterosteus aculeatus*). Other species that frequent the pond from time to time include the American eel (*Anguilla rostrata*), fourspine stickleback (*Apeltes quadrucus*), hogchoker (*Trinectes maculatus*), oyster toadfish (*Opsanus tau*), sheepshead minnow (*Cyprinidon variegatus*), striped killifish (*Fundulus majalis*), striped mullet (*Mugil cephalus*), tautog (*Tautoga onitis*), cunner (*Tautogolabrus adspersus*), white perch (*Morone americana*), and winter flounder (*Pseudopleuronectes americanus*).

Of primary abundance in the macroinvertebrate populations are the eastern (or American) oyster (*Crassostrea virginica*) and the soft clam (*Mya arenaria*). Three other bivalve mollusks – the Macoma clam (*Macoma balthica*), hard clam (*Mercenaria mercenaria*) and Morton's cockle (*Laevicardium mortoni*) – are present in much lesser numbers. The dominant gastropods are the mud dog whelk (*Nassarius obsoletus*), the common oyster drill (*Urosalpinx cinerea*) and, rarely, the thick-lipped oyster drill (*Eupleura caudata*). Of the crustaceans, the mud crab (*Neopanopeus sayi*) and the green crab (*Carcinus maenas*) are both present, although the latter occurs only rarely. Sponges include the red beard sponge (*Microciona prolifera*) and boring sponge (*Cliona celata*). Tunicates are represented primarily by the golden star tunicate (*Botryllus schlosseri*) and the sea grape, or sea squirt (*Molgula manhattensis*); this latter species is a major pest in the pond because of its tendency to form thick colonies on the culture gear, particularly buoys and nets.

The fact that many oceanic species are effectively excluded from Island Pond because of its relatively low salinity and high summer temperatures is advantageous to the oyster culture operation. The Atlantic sea mussel (*Mytilus edulis*), rock barnacle (*Balanus balanoides*) and certain species of tunicates, all of which are common to Block Island and Long Island Sounds but rarely occur in Island Pond, can be severe nuisances because of their tendency to foul nets and trays used for growing oysters. Dense mussel populations would certainly compete with oysters, not only for space but also for food. Perhaps the most serious oyster predator in Long Island Sound, the starfish (*Asterias forbesi*), has never been observed inside the pond, even though it seems likely that starfish larvae enter during high tides.

There are many salt ponds along the southern New England coastline, particularly in Rhode Island and southern Massachusetts, which are potentially productive of fish and shellfish. The fact that many are not would appear to be due to (1) legal restrictions that discourage or prohibit private control of the pond's fishery resources, (2) proximity to human development and hence vulnerability to pollution and public opposition, and (3) inability to manage the pond environment to the advantage of fish and shellfish populations. In the case of Island Pond, the first two of these potential constraints have not been factors, and, as for the third, some degree of environmental control, such as the opening or closing of the pond to the sea, is in fact possible.

It is interesting to compare the ecology of Island Pond with that of Oyster Pond on the Island of Martha's Vineyard, Massachusetts, particularly with respect to their histories of shellfish production. Oyster Pond is one of a series of salt ponds on the south shore of the Island that drain into Nantucket Sound. These ponds are relatively shallow depressions in what was once a glacial outwash plain. They formed small inlets along the coastline that were eventually closed off from the sea by a bay-mouth bar that gradually moved inshore. The two largest of these ponds, Edgartown and Tisbury Great Ponds, are roughly 540 ha in surface area and are considerably larger than Oyster Pond, with about 110 ha. Although these ponds have long supported wild shellfish populations, primarily oysters and soft clams, production has generally been erratic from year to year[1,2].

Compared with Island Pond, Oyster Pond is comparatively shallow, with a maximum depth of about 4 m when open to the sea (Fig. 1.9). The lower (south) portion is of uniform depth and is bordered to the east and west by flat terrain and to the south by a barrier beach. The northern half of the pond is narrow, shallow, and more protected from the wind by a dense growth of scrub oak. The bottom sediment around the shallow margins of the pond, out to a depth of 2 m or so, consists of a well-sorted sand-silt composition. The sediment becomes increasingly soft with increasing depth, and the bottom in the deepest parts consists of black organic ooze. The primary source of fresh water entering the pond, other than precipitation, is a narrow creek at its north end.

To sustain the natural shellfish populations in the pond, it has been customary to open it several times each year in order to maintain an adequate salt concentration. This is usually done in the spring and again in the fall, after the pond level has risen 2–3 m above sea level as a result of precipitation and inflow from small creeks. Once the

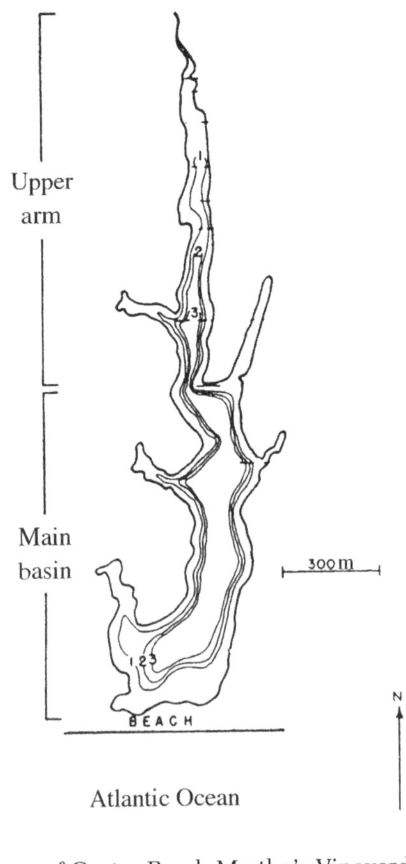

Fig. 1.9 Bottom contours of Oyster Pond, Martha's Vineyard (depth in m).

barrier beach has been breached, usually by bulldozer, the force of the outflowing water is sufficient to create a wide channel to the sea. The pond may remain open for anywhere from a week to several months, depending upon surf conditions. After it has been open for a week or more, successive inflowing tides have generally resulted in a marked rise in salinity. In the lower end of the pond nearest the opening, the salinity may reach 30‰. Once the pond closes and fresh water accumulates, the salinity begins to drop and may fall to 10‰ or even lower before the pond is opened again. Surface salinities in the larger ponds, such as the nearby Tisbury Pond, which is also opened several times each year, have a similar salinity range.

Like Island Pond, Oyster Pond tends to stratify during the summer months as the surface water warms, forming a layer over the denser, more saline bottom layer entrapped by the pond's closing. Unlike Island Pond, however, is the fact that the layer of water nearest the bottom in Oyster Pond may become warmer than the surface layer by several degrees, resulting in a positive thermal gradient[1.3] (Fig. 1.10). Both the thermocline and halocline break down during the fall, however, as stronger winds and reduction in surface water temperature result in vertical mixing, a phenomenon typical of coastal salt ponds.

The hydrography of Oyster Pond is similar in certain respects to that of the oyster (*Ostrea edulis*) polls, or pools, in Norway. Because of the high latitude, the only bodies of water along the Norwegian coast where water temperatures become warm enough to permit oysters to reproduce are the polls, which are small, deep salt water ponds that drain into fjords through narrow inlets. The oyster farmers dam the inlets in early May, allowing fresh water entering the polls from freshwater tributaries to form a lense over the denser saline water below. The lense functions as a greenhouse, allowing solar radiation to penetrate to the deeper saline layer during the day. The strong halocline prevents vertical mixing and consequent heat loss. As a result, water temperatures in the saline layer, where the oyster populations are established, rise appreciably above those in the surface layer, perhaps by as much as 10°C. This is sufficient to stimulate spawning and satisfactory larval development[1.4].

During the period 1956–1965, records on oyster spawning and setting as well as certain relevant ecological information on Oyster, Edgartown and Tisbury Ponds were maintained by government and private agencies. During this time, successful oyster reproduction occurred once in three years at best. Analysis of air and pond water temperature records established a strong correlation between the two, not surprising in view of the shallowness of the water in these ponds in relation to area. Major spawnings, usually followed by successful oyster sets, were most apt to occur in years when the pond water warmed steadily for a period of several weeks prior to the time that the temperature reached 24°C (the temperature at which oyster larvae

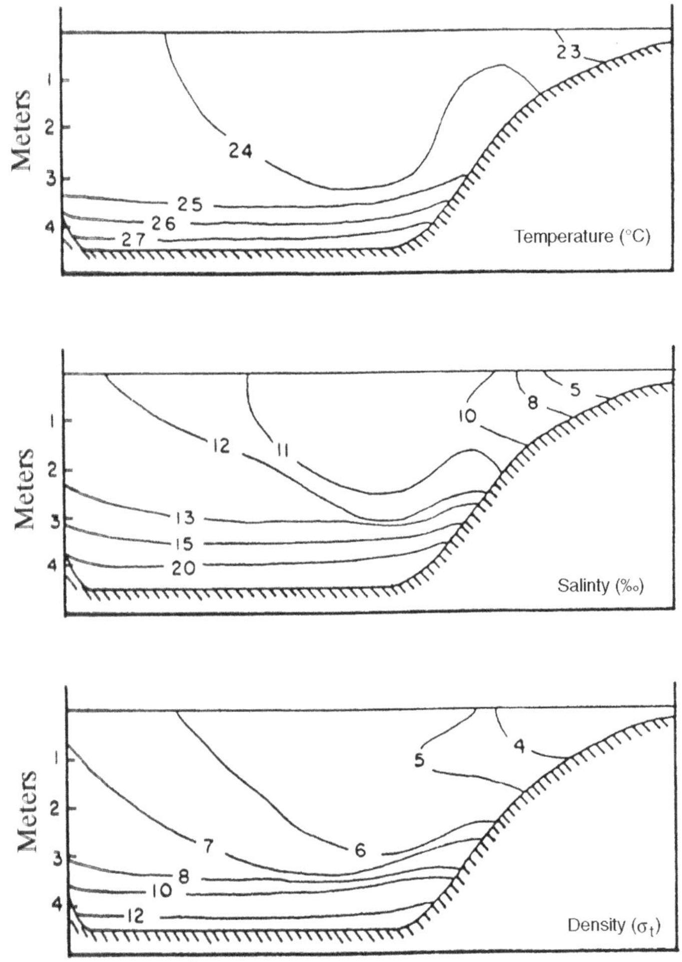

Fig. 1.10 Temperature, salinity and density distribution in Oyster Pond, Martha's Vineyard, August 1958.

were usually first detected), and when water temperatures remained at this level or higher during the larval period[1,2]. Such a temperature regime would be most likely to synchronize spawning and appeared to favor survival of the larvae. In those years in which this kind of weather pattern was disrupted by unseasonal drops in air temperature, as during periods of rain and north-east winds, there was little setting success.

It seems likely that water temperatures in these shallow ponds are insufficiently stable to ensure successful annual reproduction of oysters. Island Pond has a comparatively high depth-to-surface area ratio and is relatively sheltered from wind, resulting in a less variable temperature regime. Furthermore, unlike the barren sandy shores of the ponds on Martha's Vineyard, the rocky margins of Island Pond have afforded a favorable substrate for oysters over the years, which has encouraged a permanent natural population.

However, there are other reasons for the relative productivity of Island Pond that are more social and political than environmental. As the shore lines of the ponds on Martha's Vineyard become more and more developed, there are continual conflicts between recreational and aestheic interests on the one hand and fishing interests on the other. In recent years, residents bordering the Vineyard ponds, concerned that their cellars may flood when the ponds are high and offended by the pond vegetation when the water is low, have had a strong voice as to when the ponds are opened[1,5]. Pollution from land drainage has become more frequent. Assertion of private fishing rights over certain areas of the ponds has been opposed if not denied. In short, the socio-political climate for a sustained endeavor in shellfish culture is certainly less favorable than that experienced at Island Pond, and conflicts of this sort severely limit the potential productivity of many of the salt ponds along the coast.

2 Oyster Biology and Distribution

For many people, the life history of the oyster is a total mystery. Over the years, a good many visitors have come to see the oyster culture operation at Island Pond and have asked many questions, mostly concerning the oyster's way of life. How do they grow? How do they reproduce? What do they eat? How long do they live? Is there a male and female and, if so, how does one tell them apart? The following brief review of oyster biology and taxonomy may be helpful for those new to oyster culture.

The eastern, or American, oyster (*Crassostrea virginica*) starts out in life as a fertilized egg, microscopic in size, approximately 50 microns (μm) or two-thousandths of an inch, in diameter, that drifts about as part of the plankton. Prior to fertilization, this egg was perhaps one in a batch of 10 000 000 or more that was discharged into the surrounding water by an adult female. Occasionally, females conditioned at Island Pond for spawning in the laboratory have produced in excess of 50 000 000 eggs at one time. The act of spawning, by male and female, usually occurs in early to mid summer in New England when water temperatures have risen above 20°C and is often stimulated by the presence of the eggs and/or sperm from other oysters in the surrounding water. The probability of fertilization is therefore greatly enhanced by close proximity of the adults since the eggs are semi-buoyant and may be quickly dispersed by currents.

Within a matter of hours, the fertilized egg develops into an actively swimmimg larva. The two valves, or shells, are usually formed within 24 hours, and, by means of rapidly moving thread-like organs, or cilia, the larva is capable of propelling itself through the water. During this swimming phase, the larva feeds selectively upon minute algal cells (phytoplankton) and perhaps small particles of organic detritus that it may encounter. After a period of 10 days or so, it descends to the bottom and crawls about in search of a suitable substrate to which it may attach itself. A highly favored substrate is oyster shell, but the larvae may attach to such diverse surfaces as glass bottles, tin cans, eelgrass fronds, the top of the shell of living oyster drills, rubber tires and golf balls. At this stage, the larva is nearly 300 μm in size and is equipped with a muscular organ called a foot that enables it to move along the bottom. Once an appropriate surface is selected, the tiny oyster undergoes metamorphosis, losing its swimmimg organ and cementing itself by means of a substance secreted from a specialized byssal gland located in the foot. From this point, the oyster is no longer capable of movement and, unless somehow dislodged, remains attached to this substrate for the remainder of its life.

The European flat oyster (*Ostrea edulis*), sometimes referred to as the Belon since many of these oysters were matured for market near the mouth of the Belon River in France, has a somewhat different mode of reproduction. This species, and others of the same genus, produce relatively few eggs, perhaps a million or so, compared with members of the genus *Crassostrea*, and the eggs are considerably larger (about 150 μm in diameter). Spawning generally begins for *O. edulis* when water temperatures reach 15°C. When the eggs are extruded from the gonads, they are retained within the mantle cavity of the female rather than being discharged into the surrounding water. Here they are fertilized by sperm that have been released by nearby males and drawn into the cavity by the pumping activity of the female. The subsequent larvae are incubated for a week or more and not discharged from the mantle cavity until midway through larval development. This abbreviated planktonic existence characteristic of the *Ostrea* genus reduces the likelihood of transport by currents to unfavorable setting areas.

Oysters are primarily plankton feeders, circulating the surrounding water through an elaborate filtering system and retaining and ingesting tiny particles of organic material. The pumping mechanism is provided by the coordinated movement of fine hairs (cilia) on the gill surfaces. Particles in excess of 10 μm or so in size are usually rejected. The oyster's digestive system lacks the ability to break down cellulose and therefore the particles actually assimilated generally lack thick cellulose walls.

The growth rate of an oyster depends largely upon the amount of food in the water, the rate of current flow (a relatively strong current being beneficial in providing a constant food supply), and the surrounding water temperature. In the case of the eastern oyster (*C. virginica*), the species native to the Atlantic and Gulf coasts of North America, growth rate is usually greatest at temperatures between 20 and 30°C. The pumping rate, and hence growth rate, decline rapidly below 20°C and virtually cease at temperatures below 5°C. Under favorable growing conditions, an oyster spawned in late spring or early summer may exceed 50 mm in shell height by the end of its first growing season, i.e. mid-fall.

Oysters of the genus *Crassostrea* thrive in warm brackish water in intertidal as well as subtidal areas along the shore line, where the waters flowing from rivers and estuaries are generally rich in organic matter. Oysters of the genus *Ostrea*, on the other hand, prefer clear and cooler water of high salinity and do best in subtidal areas. The latter generally require a longer time, 4 to 5 years, to reach marketable size.

After a winter of hibernation, oysters resume feeding and growing during the spring, and, as the water warms, they begin to develop reproductive cells. Oysters are protandric, and, at this time, the difference between male and female becomes obvious. Hermaphroditism is not common[2.1], and the majority of those completing their first year are usually male. These yearling oysters develop fully viable eggs and sperm. Unless eaten by a predator, covered by silt during a storm, or infected by disease, an oyster may live for 10 years or more.

In New England, the great majority of oysters are marketed for the half-shell, or raw oyster, trade, and most of these have a minimum size (shell height or length) of 75 mm and a minimum weight of about 50 g. In most growing areas, three growing seasons are generally required to produce an oyster of this size, although certain of the faster-growing strains, cultured off-bottom, may exceed 75 mm before the end of two growing seasons, i.e. when only 18 months in age. Unfortunately for the oyster, its enemies are numerous, and the vast majority of the millions of embryos produced each year by each female oyster never reach market. In New England, the major predators, other than man, include starfish (*Asterias vulgaris* and *A. forbesi*), oyster drills (*Urosalpinx cinerea* and *Eupleura caudata*), crabs (*Carcinides maenas*, *Neopanope texana*, and *Callinectes sapidus*), whelks (*Busycon caricum* and *Busycon canaliculatum*) and, in the case of juvenile oysters, flatworms (*Stylochus ellipticus*). In addition to predators, oysters are vulnerable to periodic diseases of epidemic proportions (epizootics), violent storms and hurricanes, and deterioration in water quality as from major oil spills, industrial pollutants and toxic pesticides.

Despite these hazards, oysters manage to survive under a variety of seemingly hostile conditions. The structure of the shell enables it to withstand desiccation even if exposed to air for prolonged periods of time, a facility not shared by certain other edible bivalves, such as the bay scallop, soft clam and sea mussel, which may occupy similar habitats. Certain species such as the eastern oyster are extremely hardy animals in many respects, capable of thriving in areas with seasonal temperature extremes ranging from below freezing to 35°C and with salt concentrations ranging from near zero to over 30‰.

The majority of oysters cultured for human consumption are members of three genera in the family Ostreidae: *Crassostrea* Sacco, 1897; *Ostrea* Linne, 1758; and *Saccostrea* Dollfus and Dautzenberg, 1920[2.2]. The species *Crassostrea plicatula*, an oyster of particular commercial significance in the People's Republic of China, has been placed by some in a separate genus, *Alectryonella*, although the classification is uncertain[2.3]. Members of the genus *Crassostrea* presently account for the great majority of the oysters produced world-wide. Although pearl oysters of the genus *Pinctada*, a member of the family Pteriidae, are cultured in certain parts of the world, they are not generally consumed.

The range of the eastern oyster (*C. virginica*) extends from Labrador to Mexico, with major centers of commercial concentration in southern New England, the Middle Atlantic States (particularly the Chesapeake Bay region), and the Gulf of Mexico. The range of the Pacific oyster (*C. gigas*) is much broader. This species, referred to until recently as the Japanese oyster, is native to the western shores of the Pacific and has long been the mainstay of the oyster industries of two major producers, Japan and the Republic of Korea. It is a highly adaptable species, capable of thriving and growing rapidly under a variety of environmental conditions, and is relatively resistant to disease. In addition to supporting significant fisheries in Taiwan[2.4] and the People's Republic of China[2.5], it has been successfully introduced into New Zealand[2.6], Tasmania[2.7], Chile[2.8], the west coast of the United States and Canada[2.9], France[2.10], United Kingdom[2.11], and other countries in Northern Europe, the Mediterranean, and even certain islands in the South Pacific.

Other species in the genus *Crassostrea* contribute relatively little to total world oyster production. The Portugese oyster (*C. angulata*) at one time dominated the oyster industries in France, Portugal and Spain[2.12]; since the early 1970s, and because of disease, this oyster has virtually disappeared and has been replaced by the Pacific oyster, a similar species introduced in an effort to salvage the industry. The mangrove oyster (*C. rhizophorae*) is common throughout the islands of the Caribbean and along the shores of Central America, Columbia and Venezuela[2.13], but the oyster fisheries in these countries are quite small. Other species of *Crassostrea* include *C. brasiliana* (Brazil)[2.14], *C. tulipia* (Sierra Leone and Senegal along the west coast of Africa)[2.15], *C. iredalei* (Philippines)[2.16], *C. lugubrius* (Thailand)[2.17], and *C. belcheri* (South-East Asia)[2.18].

Within the genus *Saccostrea*, the Sydney rock oyster (*S. commercialis*) is perhaps the most important[2.19], even though restricted in its distribution to the east coast of Australia. This is a warm water species, thriving within a temperature range of 20–30°C, and supports an active fishery. A related, if not the same, species, *S. glomerata*, frequents a similar type of habitat in the North Island of New Zealand[2.20]. A third species, *S. cucullata*, occurs in India, Malaysia and other parts of South-East Asia[2.18].

In the genus *Ostrea*, the best known is the European flat oyster (*O. edulis*), the range of which extends from the Mediterranean Sea northward along the coasts of western Europe and the United Kingdom, with isolated populations in parts of Norway north of the Arctic Circle[2.21]. For commercial purposes, this species is no longer of significance throughout most of its range, having suffered catastrophic losses from overfishing, pollution and disease. It has been introduced with some success to north coastal areas of the United States, primarily Maine[2.22]. On the Pacific Coast of the United States, the Olympic oyster (*O. lurida*) is still harvested commercially in small numbers, but its abundance has been severely reduced by industrial pollution[2.23]. Most members of this genus prefer relatively cool water

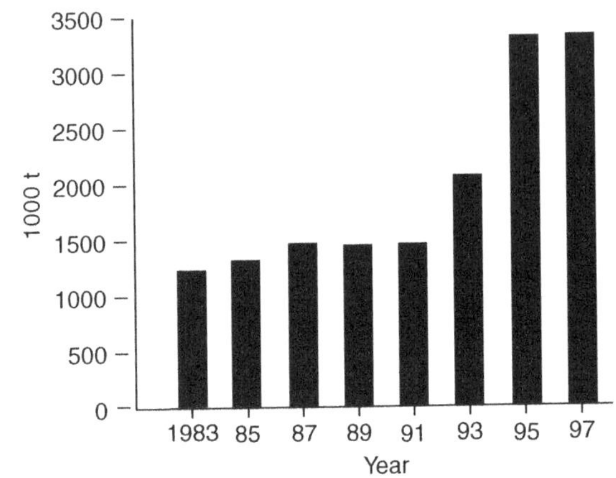

Fig. 2.1 World oyster production.

and moderate to high salinity. *O. lutaria* occurs in commercial numbers only at the southern-most tip of New Zealand, while *O. angasi* is found only in Tasmania and parts of New South Wales[2.24]. *O. chilensis* thrives at similar latitudes along the coast of Chile[2.25]. Because of the distinctive and similar larval shell structure of *Ostrea chilensis* and *O. lutaria*, it has been proposed that they be classified as the same species in a new genus *Tiostrea*[2.26]. Both produce relatively large eggs, up to 350 μm in diameter, and the larvae have a long incubation period (a month or more) followed by a very brief planktonic existence[2.27].

Since the early 1980s, the world oyster harvest has nearly tripled (Fig. 2.1), owing primarily to the enormous increase in production in China (Fig. 2.2). Historical and contemporary data on commercial oyster landings have been gathered from the following sources:

- Food and Agriculture Organization of the United Nations, Fishery Statistics, Catches and Landings (Yearbooks)
- FAO Statistical Databases, 1950–1997
- US Department of Interior, Fishery Statistics of the United States (Annual Reports)
- US Department of Commerce, NOAA, National Marine Fisheries Service, Office of Science and Technology, Fishery Statistics and Economics Database

Although no fewer than 29 species of *Crassostrea*, 43 species of *Ostrea*, and five species of *Saccostrea* have been described[2.28], only a very few of these make a significant contribution to world oyster production; the majority are too scarce, do not grow large enough to be

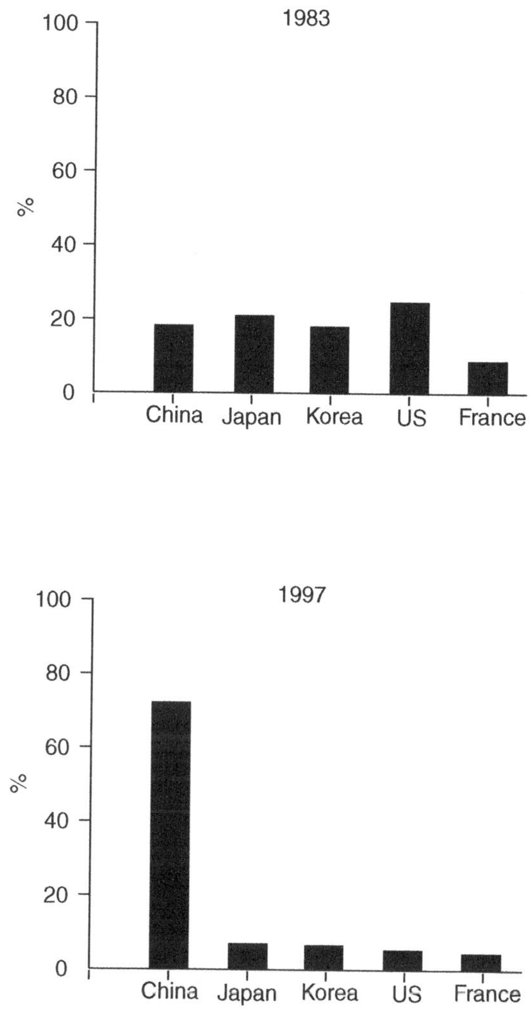

Fig. 2.2 Percentage of world oyster production by country.

of commercial interest, or, for one reason or another, are simply not exploited. The domination of the world oyster harvest by the Pacific oyster has increased steadily during the past three decades, while annual production of once prominent species such as the eastern oyster, European flat oyster and Portugese oyster have declined. In 1997, five countries accounted for more than 95% of total world production: People's Republic of China 71%; Japan 7%; Republic of Korea 7%; United States 6%; and France 5% (Fig. 2.2).

Most of the pearl oysters belong to the genus *Pinctada*, which is not included in the family Ostreidae. Members of this genus are more closely related to the mussel, attaching to the substrate by means of byssal threads, and are not valued for food. In Japan, where the majority of pearls are produced, the most important species is *Pinctada fucata*. In the more tropical regions of Southeast Asia and northern Australia, the dominant species is *P. margaritifera*.

3 Early Years

Ocean Pond Corporation was formed initially for the purpose of producing marketable oysters for the half-shell market. In 1961, the oyster industry in New England had reached about its lowest point in the century[3.1] (Fig. 3.1). The large private oyster enterprises of Narragansett Bay in Rhode Island never recovered after the 1938 hurricane, which put the finishing touches on an industry already severely weakened by inability to acquire seed oysters. In Long Island Sound, which was the heart of both Connecticut's and New York's oyster industries, efforts to recover from this hurricane's damage were impeded by repeated set failures year after year. The New England industry traditionally had relied upon the Chesapeake Bay and Delaware River as sources of seed oysters for planting[3.2], since local reproduction had become so unreliable. However, a disease caused by an unknown protozoan initially referred to as 'MSX' and eventually named *Haplosporidium nelsoni*, that first destroyed much of the oyster population in the Delaware River in the late 1950s and continued its devastation in the Chesapeake Bay several years later, effectively cut off these sources of supply. The general scarcity of oysters that prevailed in the New England region in 1961 was an inducement to utilize the potential resources of Island Pond.

It soon became clear, however, that there was little chance of producing significant quantities of mature oysters of half-shell quality from the pond. Because of the periodically low salinity of the pond water, the oysters usually had a bland taste and their inner shells were apt to be riddled with mud blisters caused by the mudworm *Polydora websteri*. Furthermore, the bottom sediment in much of the pond was too soft for oysters to thrive, and the firm ground in shallow water was not only limited in area but also infested with oyster drills. It was therefore decided to concentrate instead upon the production of seed oysters, which could be transplanted to other areas for growout and for which there was a strong demand in southern New England.

The raft culture system employed during the first year was based upon that developed and utilized in parts of Japan and tried on an experimental basis in Massachusetts[3.3]. The rafts were constructed of styrofoam logs surfaced with sheets of plywood. Boards fastened across the tops of the rafts extended over the water to provide outriggers from which oyster cultch could be suspended. Strings of cultch were constructed by threading galvanized wire through bay scallop

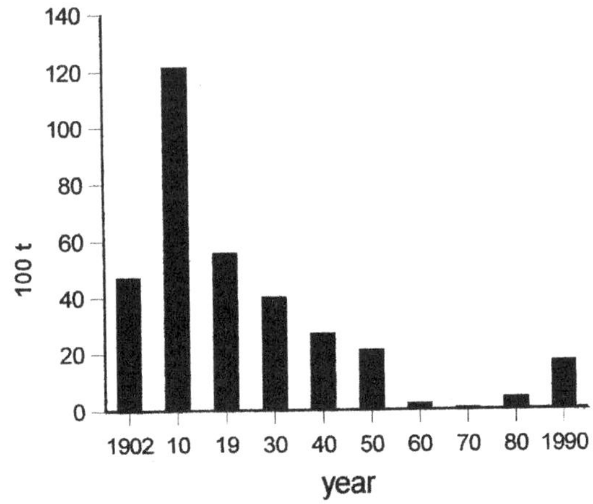

Fig. 3.1 Oyster production in New England (meat weight).

(*Argopecten irradians*) shells in which holes had been punched. Bay scallop shells were used in preference to oyster shell because they were light and easily perforated. The shells were spaced slightly apart by short (5 cm) lengths of plastic tubing in order to avoid overcrowding. The strings were then suspended from the rafts during the spawning season.

During the first year, a small number of strings were hung from the rafts in early summer. A very satisfactory set was obtained that season, leading to an expansion in the numbers of rafts and cultch strings deployed. Within a few years, the number of rafts in use had risen to 150, capable of supporting nearly 100 000 strings about 2.5 m in length[3.4].

This operation proceeded during the next 10 years as follows. Large quantities of bay scallop shell were obtained from shucking houses on Cape Cod, Martha's Vineyard or Nantucket and trucked to Fishers Island, where the shell strings were assembled. The strings were tied in bundles and stacked upon the rafts prior to the spawning season. The strings were transported to and from the rafts by means of mobile work rafts, about 2.5 m wide and 5 m long and constructed of plywood with styrofoam flotation, which were powered by small (5–6 hp) outboard motors.

Towards the end of May or in early June, the pond outlet, if open, was dammed off by means of a bulldozer in order to minimize the escape of oyster larvae. A second reason for damming was to prevent any possible intrusion of sea water that might, by lowering the pond water temperature, have an adverse effect upon spawning and duration of the larval period. It was eventually concluded that very intensive sets could occur in the pond whether it was closed or not, since tidal exchange usually was negligible and, during this time of year, a significant intrusion of sea water was highly unlikely.

During the early years of the company's operation, evidence of spawning was determined by means of oblique non-quantitative sampling, towing a plankton net that was slowly raised and lowered while under way in different parts of the pond. Plankton samples were collected several times each week, beginning in early June. Oyster larvae generally appeared in the samples when the pond surface water temperature first exceeded 23°C. The appearance of oyster larvae was usually preceded by the occurrence of the larvae of the shipworm (*Teredo navalis*) and soft clam (*Mya arenaria*). During most years, oyster reproduction was characterized by one or several significant spawnings that occurred between mid June and early July. This is considerably earlier in the year than when spawning is most apt to occur in the nearby, but much cooler, waters of Long Island Sound, i.e. mid to late summer[3.5]. When it appeared that mid to late stage larvae were the numerically dominant forms in the plankton sample, the shell strings were suspended in the water. Twice-weekly plankton sampling continued until oyster larvae were no longer detected, after which it was discontinued and the pond was reopened.

If a satisfactory set was obtained, which was most often the case, the shell strings were left suspended in the pond until the following spring, when they were harvested. This was accomplished by bringing the strings ashore, loading them on trucks, and transporting them to the main dock on Fishers Island, where they were loaded aboard oyster boats from Connecticut or Long Island. The shells bearing the juvenile oysters were stripped off the wire strings, later to be spread on private beds in New York, Connecticut and Massachusetts. At harvest time, many of the scallop shells on the strings bore more than 100 oysters each. A single year's harvest might exceed 100 000 000 oyster seed averaging about 2.5 cm in size.

It was soon determined that cultch put into the pond too early resulted in fouling of some of the shells by algae or sponge before a satisfactory set could be obtained. In order to obtain more precise estimates of larval abundance, a program of quantitative plankton sampling was initiated in 1970. Using a portable gas-powered pump, 400 liter (ℓ) samples were obtained from the surface, mid-depth and near bottom at three different locations (Fig. 3.2) in the pond twice each week, beginning the first of June. The water was passed through a plankton net, the samples were concentrated, and the number and size distribution of the oyster larvae in each sample were determined by microscopic examination. This made it possible to detect the occur-

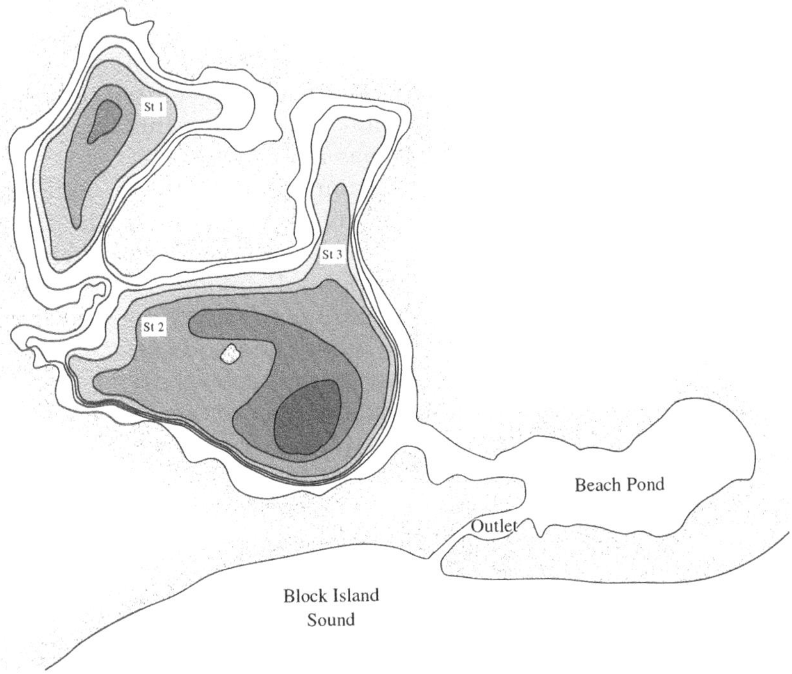

Fig. 3.2 Location of larval sampling stations.

rence and magnitude of spawning, estimate the rate of development of the larvae, and determine the abundance of larvae in relation to their horizontal and vertical distribution in the water column.

As a means of determining when spawning might be likely to occur, samples of adult oysters were obtained from the pond twice each week and analyzed for gonad condition. In 1970, for example, many of these oysters were prepared to spawn by the first of June, since their gonad layer was relatively thick and contained an abundance of active sperm and mature eggs. By mid-June, although some adults examined still had a thick gonad layer, others appeared partially spawned out. This in fact coincided with evidence, from plankton sampling, of a significant spawning that, on the basis of oyster larval size, must have occurred a few days previously.

In 1970, oyster larvae first appeared in the plankton on 8 June, primarily in the north-west sector of the pond (Station 1, Fig. 3.2). The surface water temperature in this part of the pond is usually 1–2°C warmer than in the south and east areas owing to greater protection from the wind; larvae almost always occurred in this area first. These larvae were small and thinly distributed, with maximum concentrations (about one larva per ℓ) 2 m beneath the surface (Fig. 3.3). The first significant numbers of larvae were detected three days later, particularly in the north-west sector, where replicate counts at the surface, mid-depth and bottom averaged four, three and seven larvae per ℓ, respectively. By mid-June, many of the larvae had reached setting size, and, several days later, large numbers of spat were detected on the shell strings. It seemed likely that at least some setting had occurred within 10 days after spawning.

This was followed by another major spawning in the southern and western areas of the pond (Stations 2 and 3), as indicated by the larval counts (Fig. 3.3). By 18 June, larval concentrations had reached a peak throughout the pond, with maximum concentrations of nearly 40 larvae per ℓ at depths of 2–3 m. Four days later, the majority of the larvae had disappeared from the plankton, immediately following a brief period of cold and wet weather and a sharp drop in water temperature. The fact that some of the larvae survived this temperature drop was indicated by an increase in numbers of larger larvae in the 25 June samples collected near bottom at Station 1, and the fact that setting occurred on test strings as late as 2 July.

As indicated in Fig. 3.3, the larvae were not distributed homogeneously throughout the water column. The highest concentrations of larvae were found to occur at depths between 2 and 4 m, which would be quite near the bottom at the shallowest sampling location but only at mid-depth at the deepest one. About 2 m below the surface was also observed to be the preferred depth for oyster larvae in Oyster Pond on Martha's Vineyard, where maximum concentrations were similar to those in Island Pond, i.e. about 40 larvae per ℓ[3.6]. This is not surprising, given the fact that, in the deeper areas of the pond, the bottom water was subsequently found to be quite low in oxygen and pH (Fig. 3.4). Relatively few larvae were ever found at depths greater than 4 m.

The numbers of larvae observed resulted in a very successful set during the 1970 season. Analysis of the observed larval concentrations suggest that, at least in this pond with negligible tidal exchange, a seemingly low average concentration, perhaps no more than one advanced stage larvae per liter of water, could result in a satisfactory set. A similar concentration of larvae resulting in satisfactory oyster sets have been reported for Norwegian oyster polls[3.7].

The spawning season the following year was similar in certain respects. After an initial minor spawning that was evident only in the north-west area of the pond in early June, a very significant spawning was detected 10 days later, very shortly after pond water temperatures reached nearly 24°C. The greatest concentration of larvae was initially found at Station 1, as in the preceding year. It was observed that, as this cohort developed, the larger larvae tended to avoid the surface layers and concentrate nearer the bottom at Stations 1 and 3. This is consistent with observations made from year to year, i.e. shells at the bottom of the cultch string invariably received the earliest and heaviest set; the nearer the surface, the lighter the set. The growth rate of the larvae was again observed to be rapid. It is estimated from size-

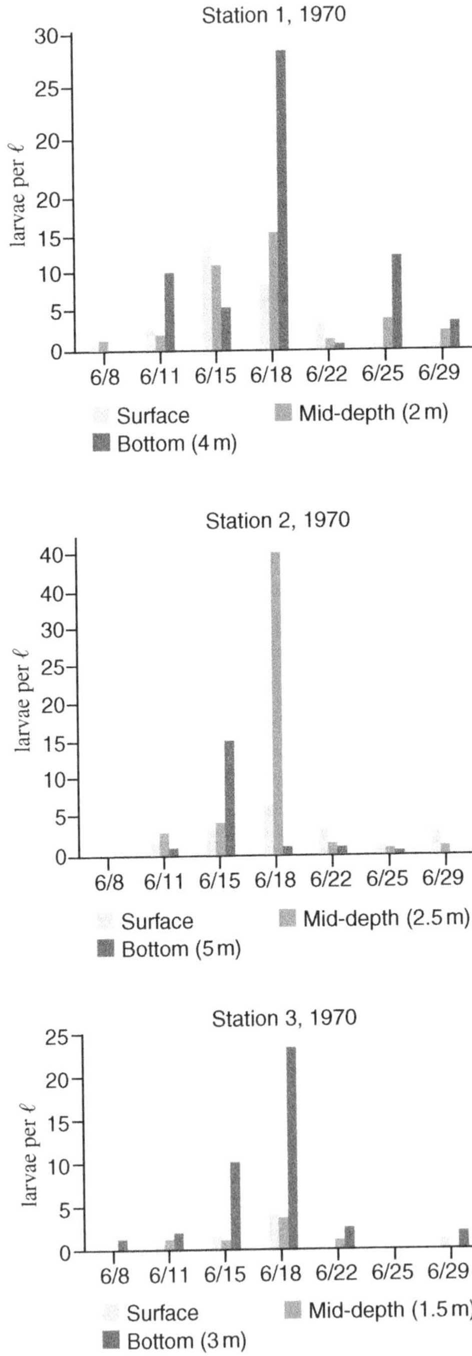

Fig. 3.3 Relative abundance of oyster larvae at different depths, June 1970.

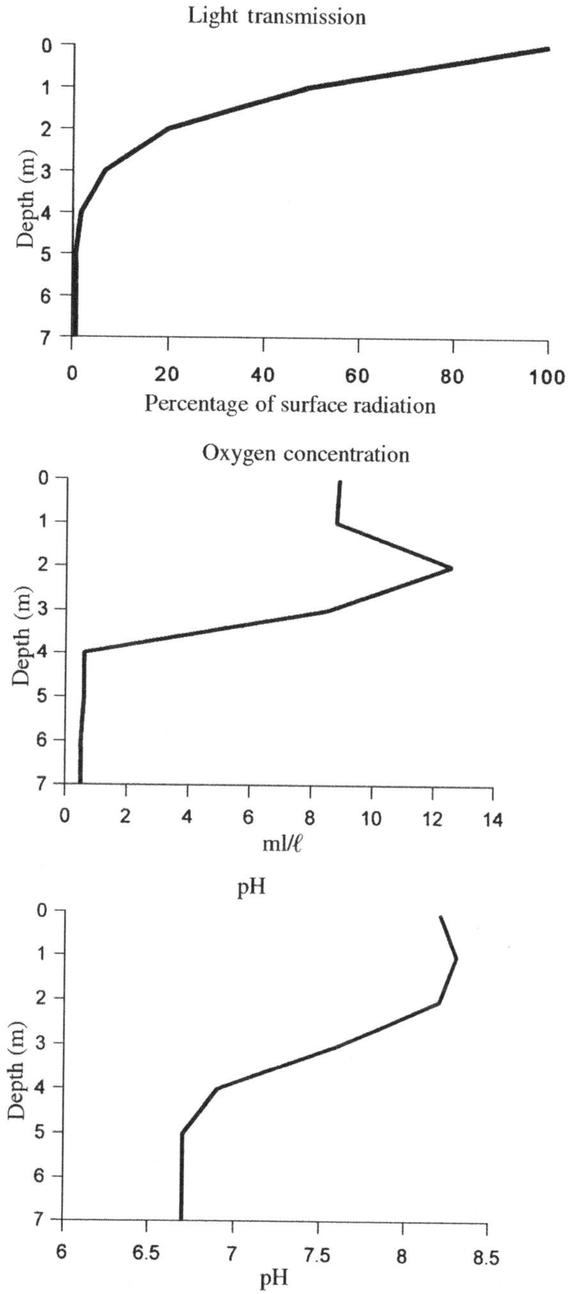

Fig. 3.4 Light transmission, oxygen concentration and pH profiles, June 1981.

frequency data that many of the larvae developed from straight hinge to setting size within a period of 8 days (Fig. 3.5). The maximum density of larvae observed, 34 per ℓ at mid-depth at Station 1, was somewhat less than the maximum observed the preceding year (40 per ℓ), but a very intensive set was obtained nevertheless.

In 1972, early-stage oyster larvae were first detected in early June, again in the north-west sector of the pond, and reached a maximum density of 27 per ℓ several days later. This was followed by a protracted period of cold, wet weather, and surface water temperatures often dropped below 20°C during June (Fig. 3.6). Probably as a result, and on the basis of larval size-frequency data, spawning was intermittent, the growth rate of the larvae was slower, and setting intensity also was observed to be low. It was apparent that, although the larvae were capable of survival and growth, albeit slow, to setting size at the prevailing temperatures, survival through metamorphosis was extremely low. Oyster larvae were observed intermittently in plankton samples through the middle of June, but rarely at densities exceeding one larva per ℓ.

In 1973, oyster larvae occurred only in small numbers during June, despite the fact that water temperatures during the first half of June were slightly above normal. After mid-June, water temperatures dropped sharply, remaining below the seasonal norm into July. In late July a spawning occurred, and, by early August, mid-size larvae were found in relatively low numbers, i.e. maximum density, five or six per ℓ at Station 1. A light set was obtained in the north-west area of the pond.

These observations suggested that oyster larvae were quite sensitive to relatively sudden drops in water temperature, during which the larvae tended to settle to the bottom. In a pond such as Island Pond, with relatively low oxygen concentrations at depths below 4 m, this pattern of behavior could be lethal for large numbers of larvae, and the success or failure of an entire season would be dependent upon prevailing weather conditions. This phenomenon has also been reported in the case of *O. edulis* in France[3.8] and *C. gigas* in Japan[3.9]. Sensitivity to unstable temperature would hold particularly true for small, semi-enclosed ponds where water temperature is closely linked to air temperature as well as solar radiation and may change relatively rapidly. In Island Pond during the years 1970–1973, major oyster sets were obtained in 1970 and 1971, but setting was only marginal in 1972 and 1973. In comparing June surface water temperatures in the north-west sector (Station 1) of the pond during these 4 years (Fig. 3.6), it is evident that temperatures rose rather steadily in the 2 years that major sets were obtained but experienced significant drops in the 2 years when setting was marginal, i.e. 1972 and 1973. As described in Chapter 1, similar results were observed in the brackish ponds on Martha's Vineyard, Massachusetts during the years 1963–1965.

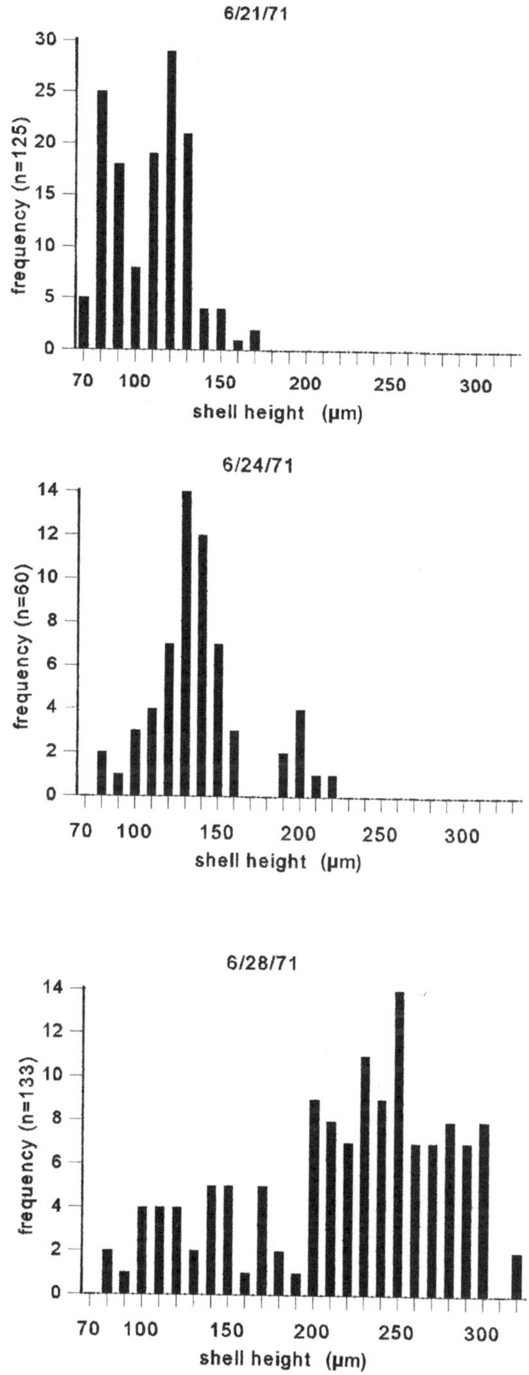

Fig. 3.5 Size-frequency distributions of larvae on successive dates, June 1971.

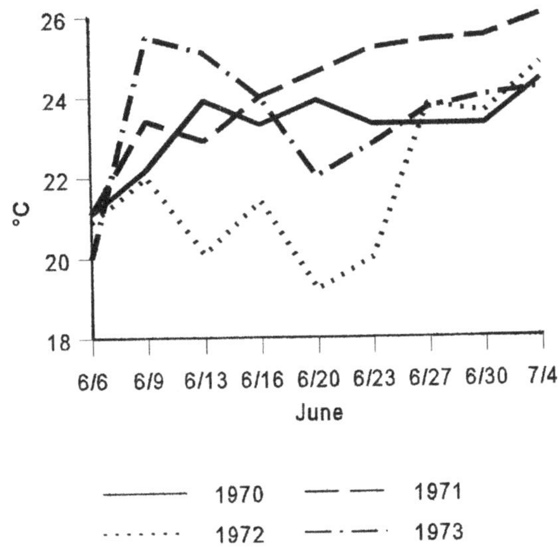

Fig. 3.6 Surface water temperatures in Island Pond during successive spawning seasons.

This element of uncertainty in oyster reproduction under natural conditions was one factor leading to the temporary discontinuation of oyster culture in Island Pond. Another factor was the inability to control setting intensity. Severe overcrowding of oyster spat on the shells, which was frequently the case, was discouraging to the buyers, since this could result in high mortalities, and the survivors were often stunted and misshapen. A third factor was the occurrence of a series of excellent oyster sets in Long Island Sound during the late 1960s and early 1970s, slackening the demand for seed oysters. Finally, by the 1970s, hatchery production had become more reliable, offering certain advantages, such as choice of parent stock and control over setting intensity, that theretofore had not been available. Therefore, oyster culture efforts in Island Pond were discontinued after 1973 and not resumed until 1977.

4 Oyster Culture in the Far East

According to fishery statistics compiled by the Food and Agriculture Organization of the United Nations, the United States accounted for nearly 60% of the world's oyster harvest in the early 1950s. During the next 45 years, the United States percentage of world landings steadily declined, falling to 6% by 1995. This shift in global production was the result not only of a decline in United States production, which has been severe, but also of a remarkable increase in production in other areas of the world, most particularly in the Far East. Between 1950 and 1995, world landings increased by nearly a factor of ten, from 418 000 to 3 266 000 metric tons (tonnes, t). Presently 85% of the world harvest can be attributed to Japan, South Korea, and the People's Republic of China. The following account of oyster culture methods in these countries is based upon discussions with shellfish biologists and growers, personal observation, and a review of the pertinent literature. Here, as in subsequent chapters dealing with oyster landing statistics, use has been made of the Statistical Databases of the Food and Agriculture Organization of the United Nations.

In Japan, an intensive form of oyster culture has been pursued for several centuries. Beginning here in the seventeenth century, oyster culture consisted primarily of enclosing small areas of intertidal flats with bamboo fencing, which provided some protection from predators and siltation. Within the enclosed area, bamboo stakes were imbedded in the bottom sediment in rows at the beginning of the spawning season, providing a suitable surface for the progeny of wild oysters to attach and grow[4.1]. Rocks and oyster shell were also spread on the flats for this purpose. At the end of the second growing season, the oysters were simply knocked off the stakes and the larger ones marketed. Those too small for market were spread on the bottom for additional growth.

This method of culture persisted until the 1920s, when the 'hanging culture' method was first initiated in Japan[4.2]. Initial experiments showed that growing oysters off the bottom, by *suspension*, had the advantages of (1) improving growth rate, since the oysters were submerged at all times and could feed throughout the tidal cycle, (2) greatly reducing mortality, since bottom-crawling predators were now unable to reach their prey, and (3) offering greater flexibility and expansion in the use of growing areas, since depth and nature of the bottom would not be a major consideration. The development of off-

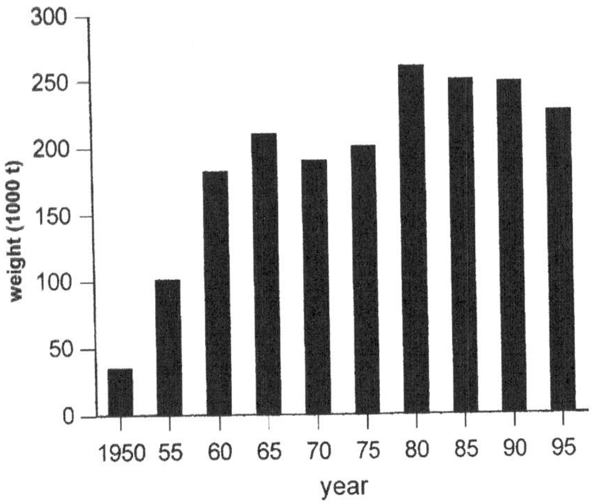

Fig. 4.1 Oyster production (whole weight) in Japan.

bottom culture essentially revolutionized the Japanese oyster industry, leading to a rapid increase in production that was soon comparable with that of the United States (Fig. 4.1).

Hiroshima is at the heart of the Japanese oyster industry. Here, in the Seto Inland Sea, nearly 80% of Japan's oyster production takes place. (Miyagi Prefecture accounts for about 10% of total production of adult oysters as well as the majority of seed oysters distributed to other growing areas or exported overseas. Okayama Prefecture accounts for another 10%). Osamu Fukuhara, a biologist at the Nansei Regional Fisheries Research Laboratory, and Dr Tomohiro Kimura, Director of the Hiroshima Prefectural Fisheries Experimental Station, were particularly helpful and informative about the industry in Hiroshima and Japan in general, and arranged visits to local oyster operations.

Although two species, *C. gigas* and *C. rivularis*, are cultured for consumption in Japan, the former is much the more widespread and is the basis of the nation's industry. Natural populations of this species occur as far south as Taiwan and Hong Kong[4.3] and as far north as Sakhalin Island in Russia[4.4].

Several factors contribute to the remarkably high yields achieved in Japan. The first is that the Pacific oyster is a rapid grower when compared with, for example, the American oyster. Although the majority of oysters produced in Japan are 16–20 months old at the time of harvest, a significant number may be less than 1 year old. The second factor is climate. In Hiroshima Bay, water temperatures rarely fall below 10°C, allowing the oysters to grow and fatten the year around. The third, and perhaps most important, factor has been the development of the hanging culture technique.

Another contributing factor has been the commitment of the government to the development of aquaculture. This is reflected by the relatively modest rents that oyster farmers must pay to the government for the use of growing grounds, and by the impressive amount of research undertaken by government-funded agencies to assist these farmers.

Oyster production in Japan has achieved some of the highest yields per unit area of any part of the world. As a result of hanging culture techniques, production of oyster meat from long-lines and rafts may be as high as 20 000 kg/ha per year[4.5]. However, these oysters are cultured not so much as a luxury food as an important component of the national diet as well as an exportable commodity. Since the oysters are shucked immediately upon harvest, there is little concern about the size, shape or quality of the shell, important features in the half-shell industries of western Europe and the United States.

The ancient bamboo stake method is still utilized in certain calm and sheltered areas, free of pollution, where the soft mud bottom would prevent other forms of bottom culture. Bottom sowing, which consists simply of spreading immature oysters on intertidal bottom, may also be practiced where the bottom is firm enough to prevent the oysters from sinking into the sediment. Bottom grounds are sometimes improved by adding stones, lengths of bamboo, shells, or other materials to provide a firmer substrate. However, the great majority of Japan's oysters are cultured off the bottom, suspended from racks, rafts or long-lines[4.6,4.7].

Intertidal racks are commonly used for collecting the juvenile oysters, or spat, during the spawning season (Fig. 4.2). Holes are punched through oyster or scallop shells, which are then threaded through with galvanized wire to form strings. The shells are spaced a few cm apart by short lengths of plastic or bamboo tubing. When spawning begins and sampling by government biologists indicate a satisfactory number of larvae in the water, the strings are hung from the racks. The timing for this is rather critical. If hung too early during the spawning season, the shell may become silted or fouled, particularly by barnacles. If hung too late, the density of spat may be too low.

Racks are also used in the 'hardening' process. This is designed to eliminate weak seed oysters and increase the hardiness of the survivors, some of which may be intended for overseas shipment. The process involves exposing the juvenile oysters to air for 6–10 hours each day. Strings initially hung from the racks to collect spat may be allowed to remain on the racks for several months for hardening, or strings that were suspended from rafts to collect spat may be brought back ashore, several months after setting has occurred, for hardening. The process usually begins in September and may continue until the following spring. Oysters that settled on string collectors placed on intertidal racks have already been subjected to some hardening and are often moved from the racks to the growing rafts or long-lines late in the fall of their first year rather than the following spring (Fig. 4.3).

Fig. 4.2 Racks for collecting oyster spat, Hiroshima Bay, Japan (photo by O. H. Matthiessen).

Fig. 4.3 Oyster growing racks, with rafts in background, Hiroshima Bay, Japan (photo by O. H. Matthiessen).

In the Hiroshima Prefecture, the practice of stick and bottom cultivation has all but disappeared, and racks are used only for the purposes described above[4.8]. For growth to marketable size, as well as for fattening purposes, rafts are used almost exclusively in this part of Japan. The rafts, consisting of frames of bamboo logs supported on floats (tarred oil drums or styrofoam logs), measure about 10 m in width and 20 m in length. The rafts are usually tied together in series of five and moored at each end in areas considered most favorable for growth, i.e. where the water is deep and the current swift.

Before the juvenile oysters are suspended from the rafts, however, the original strings used for collecting the spat are disassembled, and the shells bearing the seed are restrung on heavier wire in lengths of about 10 m. During the restringing, the shells are separated by longer spacers, 15 cm or so in length, in order to prevent overcrowding as the oysters grow. Each raft may suspend anywhere from 650 to 780 strings each. The density of juvenile oysters on the shells will vary depending upon the intensity of the set and the survival during the hardening process but can be as high as 40–50 oysters per shell. Because of competition by the oysters for both food and space, the optimum number is considered to be about 15 per shell.

In order to avoid mortality during summer, some oyster farmers begin to harvest the oysters during the winter of their first year, when the oysters are only 6–9 months of age. It is more usual, however, to follow the 2 year cycle whereby the oysters are hardened for a period of 7–9 months or so, beginning in the September of their first year, and are then allowed to grow and fatten during the following summer and fall. Under this schedule, the oysters are usually 16–21 months of age when harvested.

Because oyster cultivation by rafts in Hiroshima Bay is so intensive, seasonal reductions in dissolved oxygen have resulted in significant oyster mortalities[4.9]. It has been calculated that annual waste production from oysters suspended from a single raft ranges between 0.6 and 1.0 t dry weight. This fecal material is high in organic matter and causes anaerobic conditions beneath the oyster rafts. In anticipation of this, many of the oyster farmers tow their rafts from Hiroshima Bay to the outer reaches of the Inland Sea in early summer in order to avoid low oxygen concentrations.

It has also been found that the location and spacing of the rafts, and the number of strings suspended from each raft, can be important, since the food concentration in the water passing through a series of rafts is depleted by the oysters rather quickly. The yield per raft in Hiroshima Bay may be reduced substantially if the number of rafts is not controlled[4.8,4.10].

While the raft method of culture has been extremely successful in Japan, it is not likely to be adopted in the United States. One reason

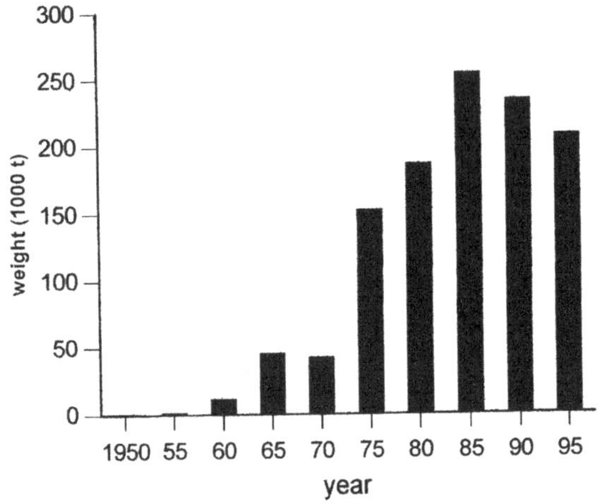

Fig. 4.5 Oyster production (whole weight) in the Republic of Korea.

shore line of the west coast. At about the time that the production curve in Japan was beginning to flatten out, Korean production began to climb rapidly, increasing by a factor of nearly ten during the next 20 years. This remarkable growth was largely the result of active government participation and support, and the adoption of the Japanese hanging culture techniques[4.12]. This method was found to be particularly suitable in the numerous sheltered bays along Korea's southern coastline.

In Taiwan, the development of oyster culture has been even more recent. The native oyster is the Pacific oyster (*C. gigas*). Since 1980, annual production has averaged around 25 000 t. Until the 1970s, most production came from stick culture in the intertidal zone, using bamboo stakes[4.13]. The stakes, nearly 1 m in length, were imbedded in the bottom sediment during the spawning season. Often strings of oyster shells were strung from stake to stake to provide additional, and probably better, cultch material. After setting, the oysters were allowed to remain on the stakes until they reached marketable size, at which time they were knocked off and shucked for market. Because of the relatively high water temperatures, some of the oysters would reach harvestable size within 4 months.

Increasing rates of intertidal land reclamation and pollution have made it necessary for the oyster growers to move into deeper, cleaner water farther from shore. Following the experience of Japan and Korea, the great majority of oysters produced on Taiwan today are grown from rafts, racks and long-lines. Intertidal racks are used to collect spat[4.14]. Water temperatures, ranging from 12 to 30°C, favor rapid growth, and most oysters are harvested within one year.

The country in which oyster production has shown the most extraordinary increase in recent years is the People's Republic of China.

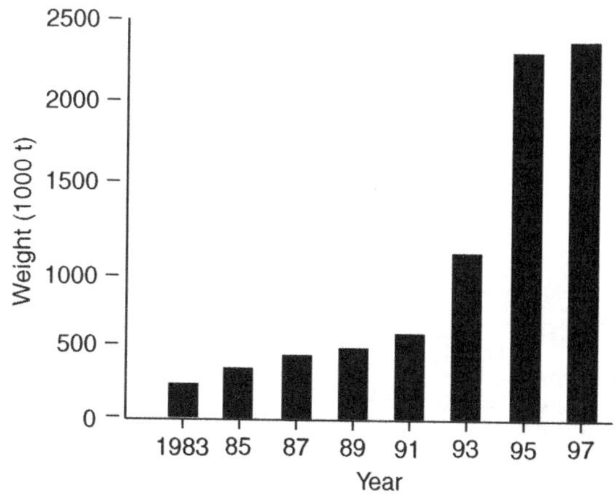

Fig. 4.6 Oyster production (whole weight) in the People's Republic of China.

Oyster culture in China is said to have begun about 2000 years ago during the Han Dynasty[4.15]. What little has been reported about oyster culture in China suggests that, until relatively recently, methods changed very little over the centuries. Between 1983 and 1997, however, oyster production experienced over a tenfold increase[4.16] (Fig. 4.6), making it the leader in world production. Seafood is a major component in the Chinese diet, and the majority of oysters are sold for domestic consumption.

There is some uncertainty as to the identity of the species being cultured in China. Although more than 17 different species of oysters have been described[4.16], some are considered to be regional varieties of *C. gigas*. The most important in order of volume of production are the zhe oyster, *Crassostrea plicatula*, cultured primarily along China's south coast; the Suminoe oyster, *C. rivularis*, found in most of China's estuaries; and the Pacific oyster, *C. gigas*, which was introduced from Japan[4.17].

In some parts of China, the traditional intertidal methods of farming are still pursued (Fig. 4.7), although the cultch materials for collecting spat – stones, pieces of ceramic, tiles and oyster shells – have been replaced by cement bars[4.18] (Fig. 4.8). These are about 0.5 m in length and are impregnated with bamboo for reinforcement. The bars are inserted vertically into the mud and have the advantage over the traditional types of cultch in providing greater surface area for spat attachment. Furthermore, they do not sink into the mud and offer the attached oysters slightly more protection from bottom-crawling predators. After setting, the oysters are allowed to grow on the bars for up to 4 years before harvesting. Another form of intertidal culture is the use of racks constructed of concrete. Cultch strings, consisting of wire threaded through oyster shells or tiles, are suspended from the racks.

Fig. 4.7 Fishing village, Laufaushan, China (photo by O. H. Matthiessen).

Fig. 4.8 Cement stakes used as cultch in Laufaushan, China (photo by O. H. Matthiessen).

Fig. 4.9 Carrying oysters ashore in Laufaushan, China (photo by O. H. Matthiessen).

The oysters are harvested from the bars either by tongs during high tides or by hand when the flats are exposed at low tide (Fig. 4.9). Very often a device called a wooden horse, a flat board upon which the oyster farmer stands with one foot and propels himself with the other, is used to cross the soft mud flats. For fattening purposes, the oysters are placed in baskets, which are suspended beneath rafts moored in high salinity water, for a period of 60 days or more, after which the oysters are brought ashore and shucked. The shucked meats are sold fresh, dried or canned.

Stake culture is still practiced in Fujian province[4.16], using 1 m lengths of bamboo. In order to condition the stakes for setting, they are first buried in sand for 3 months or dried in the shade. They are then stacked in bundles in the intertidal zone in order to attract a set of barnacles. After a few months, the stakes are brought ashore, the barnacles removed, and the stakes allowed to dry in the sun. The calcareous attachment discs of the barnacles form an attractive site for the settlement of oyster spat. The stakes are then inserted into the bottom sediment. During the growth of the oysters after setting, the stakes may be moved periodically, usually to positions lower in the intertidal zone in order to increase the feeding period.

In Guangdong Province, hanging culture methods, using rafts made of bamboo and styrofoam, were introduced in the late 1970s. The cultch often used consists of cement-coated asbestos boards or oyster shell suspended on wire. Also used is cement-coated rope. Oysters may be fattened by removing them from the original cultch and holding them in cylindrical net cages.

In recent years, the raft and long-line methods have become generally adopted in China, since both growth rate and quality of the oyster are very much superior to those obtained by the traditonal bottom-culture methods. In addition, even though the majority of oysters produced derive from natural reproduction, oyster hatcheries have come to play a significant role in the industry[4.17]. Oyster spat are collected on strings of shell suspended in large concrete tanks, and the seed are grown to maturity in suspension or on the bottom.

Factors that favor a high level of oyster production in China include an extensive coast line, about 18 000 km, with a large number of estuarine areas suited for both bottom and suspension culture, an environment that allows many of the oysters to reach marketable size within a year's time, and a strong public demand for seafood[4.16]. However, after the extraordinary growth in the oyster fishery since 1990, annual production for 1995, 1996 and 1997 appears to have leveled off, at least temporarily, at about 2 300 000 t.

5 Developments in Culture Techniques

During the 1970s, methods of rearing oysters in commercial hatcheries and research laboratories became more dependable. This was due in large part to the pioneering work of Dr Victor Loosanoff and his colleagues at the National Marine Fisheries Service Shellfish Research Laboratory in Milford, Connecticut, during the 1950s and 1960s. One of the more useful innovations in hatchery technique, initiated by Bill Budge of Pacific Mariculture in Pescadero, California, involved the use of 'micro' cultch as an alternative to whole oyster shell. By grinding the shell down to a fine size, 200–350 µm or so, or about the same size as a late-stage larva, it was possible to produce so-called 'cultchless', or single, oysters, that were free of clusters. This resulted in a well-shaped oyster best suited for the half-shell market. For many growers, faced with the problem of culling overcrowded clusters of misshapen oysters, single oysters were much to be preferred over the natural sets obtained from Island Pond.

Ocean Pond Corporation had temporarily suspended operations in 1973, but advances in culture techniques such as these revived interest in oyster farming in New England, creating renewed demand for seed oysters. In 1977, a small sample of cultchless seed, 2–3 mm in size, was obtained from Ian Walker, an oyster grower in Maine. Adopting his 'nursery culture' method, the seed were placed in several floating trays in Island Pond. These trays consisted of rectangular wooden frames with fly screen top and bottom, which were moored in the Pond in early summer. Every few days, the trays were flipped over to expose the underside to the air and sun and thereby prevent clogging of the mesh by fouling. By the end of the summer, the oysters had experienced no mortality, ranged from 25 to 50 mm in shell height and were considered large enough to plant for growout on the beds of the Cotuit Oyster Company on Cape Cod, Massachusetts. Located on Cotuit Bay, the Cotuit Oyster Company produced market-size oysters of high quality, but, since natural oyster sets rarely occurred in Cotuit Bay, depended upon other sources of seed.

These initial results were sufficiently positive to justify enlarging the system the following year. In late April, about 2 000 000 tiny seed, set on hard clam (*Mercenaria mercenaria*) shell fragments, were obtained from the hatchery of F. M. Flower and Son Oyster Company in Bayville, New York. These were placed in fly screen trays at an intial density of about 40 000 oysters per tray. The trays measured 0.6 × 1.2 m and were about 10 cm deep. The wooden frames had been dipped in a non-toxic vinyl paint to discourage damage by shipworms. The trays were tied end-to-end and arranged in a series of parallel lines, about 180 m long and 6 m apart, which were moored at each end in the western section of the pond. The trays were flipped over three times each week.

By mid-summer, herring and black-backed gulls, finding the trays convenient perches, had begun to shred the plastic fly mesh in their efforts to eat the grass shrimp and amphipods that had taken residence in the trays. The gulls appeared to have no interest in the oysters but were bent on destruction. Covering the trays with heavy-duty quarter-inch (0.6 cm) plastic mesh eventually proved to be a satisfactory deterrent, but not before many of the trays had been badly damaged.

The decision to use surface trays was based on the fact that the bottom of the pond was, for the most part, too soft for the survival and growth of small oysters, as described earlier, and the limited amount of firm bottom was heavily populated with predatory oyster drills (*Urosalpinx cinerea*). Furthermore, during the preceding year, mortality rate in the experimental trays appeared to be negligible, and, despite the absence of tidal currents in the pond, the rate of growth was surprisingly favorable.

Since there had been no previous experience with surface tray culture in the pond, a series of test trays were also placed in the pond to determine the optimum density for satisfactory growth. The test densities selected were 5000, 10 000, 20 000, 40 000 and 60 000 per tray. Comparative growth rates are shown in Fig. 5.1. It became evident that, up to a size of about 8 mm – the size at which the oysters would

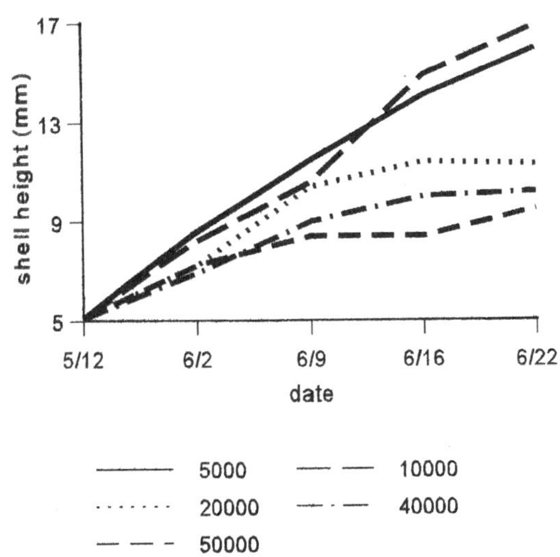

Fig. 5.1 Comparative growth rates of juvenile oysters held in 2 mm mesh trays at different densities.

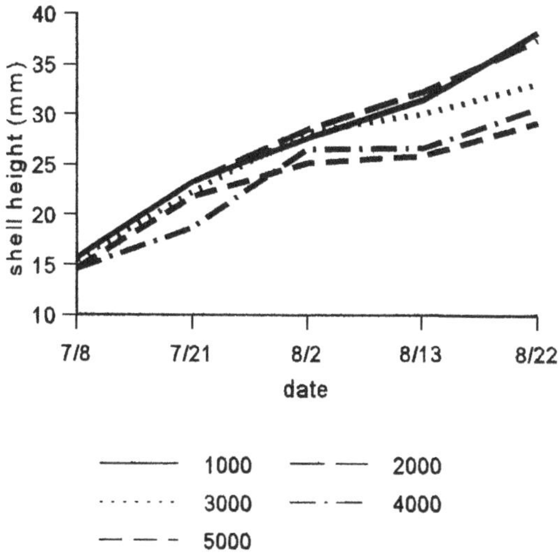

Fig. 5.2 Comparative growth rates of juvenile oysters held in 6 mm mesh trays at different densities.

be transferred to trays of larger mesh – growth rate at densities of 20 000 per tray was about the same as that at lower densities but superior to the growth rate observed at higher densities. Therefore the tray density for juvenile oysters in this size range was standardized at 20 000 per tray, or about three per cm^2 of surface area.

By early June, the oysters had grown large enough to begin overcrowding the trays and were then sieved in a 6 mm mesh screen. The slower growers that passed through the sieve were returned to fly screen trays; those retained were placed in trays having 6 mm galvanized wire mesh, similar in dimensions to the fly screen trays, at an initial density of 4000 per tray. The frames of these trays were also constructed of wood dipped in vinyl paint. Styrofoam blocks were inserted at each end of the tray to provide additional buoyancy. For purposes of establishing the optimum density for growth, additional oysters were held in test trays at densities ranging from 1000 to 5000 per tray. It was determined by mid-August that a density of 2000 per tray, or about 3000 oysters per m^2 of surface area, was most suitable for oysters in this size range, since growth at this density was nearly the same as that with 1000 per tray, and superior to those at higher densities (Fig. 5.2).

By mid-August, a significant number of oysters had exceeded 25 mm in size and were considered large enough to transfer to the beds in Cotuit. Oysters below this size are extremely vulnerable to bottom predators, particularly crabs. Oysters that were retained in a 25 mm mesh sieve were shipped to Cotuit; those that passed through were returned to the 6 mm mesh trays for further growth. The sieving process continued until mid-October, when shell growth had nearly terminated for the season.

Since the surface trays would be vulnerable both to strong winds and ice during winter, it was necessary in the fall to transfer the oysters retained in the pond to non-buoyant plastic (Nestier) trays, about 0.6 × 0.6 m in dimension and designed to be stacked on top of one another. These trays were filled with about 1000 oysters each, stacked in bundles of ten trays, and suspended beneath the surface from buoys. Here the oysters remained until the following spring, when they were removed from the plastic trays and returned to floating trays for additional growth. By about mid-June, the last of the 1978 year-class were large enough to be transferred to Cotuit. A total of nearly 2 000 000 oysters of this year-class were ultimately spread on the Cotuit beds.

In 1979, it was possible to obtain hatchery-reared oyster seed from Maine, and essentially the same rearing system was followed as in 1978. However, the availability of seed oysters from regional hatcheries was inconsistent. Very rarely were orders completely filled, and, in some years, the supply of seed was negligible. Also, some batches of imported seed gave very poor results in terms of survival and growth. For this reason, Ocean Pond Corporation initiated its own hatchery system to supplement the shipments from other hatcheries. This project began as an experiment, initially intended as an emergency back-up, with as little expense involved in its construction and operation as possible, but eventually became the primary source of supply.

Initial efforts to produce hatchery-reared seed oysters at the pond site were begun on a trial basis in 1980. The original hatchery was essentially a small greenhouse attached to a pre-existing and equally small storage shed. It consisted of a wooden frame structure with plywood floor, measuring about 2.5 × 3 × 2.5 m high, enclosed with clear fiberglass panels. It was large enough to accommodate ten 160 ℓ cylindrical plastic containers, which served as larval tanks. Pond water was supplied to the hatchery through a garden hose connected to a 0.5 hp electrical pump. The entire structure was disassembled each fall to avoid wind damage during winter.

When the pond temperature in late spring has risen to 21°C, the majority of the adult oysters in the pond are ripe and ready to spawn. Large adults showing new and prominent shell growth were collected from submerged rocks near the hatchery in early June, cleaned of fouling organisms, and placed in pyrex dishes filled with filtered pond water. The water temperature in the dishes was slowly elevated to 30°C by placing the dishes in the sunlight on top of sheets of black plastic. The oysters were then stimulated to spawn by macerating one of the adults in a Waring blender and adding a sperm or egg suspension to the dishes. In our experience, attempting to obtain viable embryos by stripping the eggs or sperm from adults of this species, rather than waiting for the oysters to release their gametes naturally, has not produced favorable results in terms of larval survival. The eggs obtained from the spawners were then mixed in a bucket with a small quantity of sperm, and the embryos transferred to the larval containers. After 24 hours, the majority of the larvae reached the straight-hinge stage, and their density in the tanks was adjusted to about 30 000 larvae per ℓ, or nearly 5 000 000 larvae per container. From this point on, the water in each larval tank was changed each day by draining the water through a Nitex sieve of small enough mesh size to retain the larvae, cleaning and refilling each tank with filtered water, adding algal food and then replacing the larvae.

Algal food for the larvae was obtained by fertilizing a series of 100 gallon outdoor tanks each day with nutrient-rich water pumped from the bottom of the pond and brought to shore in large refuse containers. About 360 ℓ of pond surface water were pumped through a bag filter into the tanks, to which was added about 40 ℓ of bottom water. Under fair weather conditions, a phytoplankton bloom consisting primarily of microflagellates and small diatoms would occur in the tanks within a few days, providing a satisfactory food for the larvae. These cultures would generally ‘crash’, i.e. deteriorate, a few days after peaking, and it was expedient to inoculate another tank with about 40 ℓ from the initial culture before it crashed. Under favorable weather conditions, it was possible to maintain and harvest a series of cultures before having to repeat the fertilization procedure.

These cultures would usually peak at cell densities of about 2 000 000 cells per ml. To achieve the desired food concentration in the larval tanks of about 100 million cells per ℓ, about 4–8 ℓ of food were added to each tank each day.

In this system, many of the larvae would attain setting size of about 275–300 µm 10 days or so after fertilization. As the larvae reached this size, they were placed in shallow trays with fine mesh bottoms, in which fine particles of ground oyster or bay scallop shell had been spread as cultch. The density of the larvae in the setting trays, 0.6 m × 0.6 m × 10 cm deep in dimension, was about 100 000 per tray. The cultch particles were just large enough to be retained in standard plastic fly screen. Dolomite particles of similar size, used as aquarium filter material and obtainable at most pet stores, were also used as cultch with satisfactory results. The trays, which were buoyant, were placed near shore in shallow water. After 48–72 hours, the shell particles with spat attached were transferred to the standard fly screen trays, described earlier, which were then placed out in the pond for further growth.

This was a primitive system to say the least, justified only by its simplicity and minimal expense, and production depended heavily upon prevailing weather conditions. Extended periods of cold and cloudy weather limited the algal food supply by delaying the development of the nutrient-rich bottom layer, which depended upon thermal stratification. Poor weather also discouraged algal blooms in the outdoor tanks and slowed the development and metamorphosis of the larvae in their unheated containers. Nevertheless, barring unfavorable weather, it not only proved to be a valuable supplement to the numbers of seed oysters obtained from other hatcheries but in some seasons was the only source of supply.

During the past 20 years of operation, our system of oyster production has undergone a series of changes. The use of 6 mm mesh floating trays was eventually abandoned in favor of pearl nets. The survival rate in the trays was generally excellent, as was the rate of growth. However, during the course of the season, the trays became waterlogged and increasingly difficult to handle and, as a result of hard use and deterioration, required constant maintenance, including replacing torn mesh and repairing broken frames. During late summer, the trays often became choked with green algae (*Ulva lactua*). Algal fouling could be controlled by covering each tray with plywood boards to block the sun, but the boards had to be removed in order to flip the trays, and, with more than 500 trays in use at one time, a significant amount of labor was required. Periodic storms and the occasional hurricane wreaked havoc with the gear; inevitably lines of trays would break apart and end up hopelessly entangled in various parts of the pond. In some instances, it was necessary to retrieve trays from the surrounding bushes, many damaged beyond repair and with their oysters lost.

Perhaps the most important reason for shifting to pearl nets was the anticipated reduction in labor required to operate the system. In addition to maintenance it was necessary with trays to 'winterize' and 'dewinterize' the oysters, transferring them from surface trays to submersible trays in the fall in order to avoid ice and winter storms, and returning them to the surface trays in the spring if additional growth was required. By shifting to pearl nets, this procedure was eliminated, since the nets, suspended from buoyed long-lines by tub-trawl clips, are positioned a minimum of 16 cm below the surface. The current procedure is to transfer oysters from the surface fly screen trays to the nets as soon as they reach about 6–8 mm in size, at densities of about 250 oysters per net. Three nets are fastened together in series, one below the other, and the top net is clipped to the long-line. These lines, comprising 8 mm polypropylene about 350 m in length, are moored in the south section of the pond. Once the oysters go into the pearl nets, they are never removed until they are harvested. This gear arrangement has survived well, even under hurricane wind conditions.

The pearl nets are remarkably durable, good for 3–5 years of hard use in our experience, and relatively inexpensive. Obviously, a good many nets, nearly 8000, are required to accommodate current annual production of nearly 2 000 000 oysters, and the amount of time

Fig. 5.3 Pearl nets being air-dried to reduce fouling (photo by William Greenough).

required for just loading and tying, about 70 man-hours during the season, is significant. The major drawback, and the one afflicting all suspension culture systems to a greater or lesser extent, is that the nets are subject to intensive biofouling. One of the advantages to culture in Island Pond is the absence of many fouling organisms, particularly the mussel (*M. edulis*) and rock barnacle (*Balanus balanoides*), capable of virtually prohibiting off-bottom culture in certain areas. However, the major fouling organism in Island Pond is the sea squirt, a tunicate, *Molgula manhattensis*, which can virtually cover the nets in late summer and early fall, to the extent that the nets can scarcely be lifted from the water. If left untended, circulation through the nets is reduced if not eliminated. Fouling is controlled by air-drying the nets on warm, sunny days (Fig. 5.3) or jetting them by pump. Dipping the nets in a saturated salt solution before air-drying appears to offer little advantage. During particularly warm seasons, however, when tunicate fouling is usually most severe, anti-fouling measures must begin by July and intensify during the remainder of the summer and into the fall. During September and October, net maintenance can become a full-time occupation for three people. The normal work force at Island Pond, from mid-March through October, is two or three people, depending upon the amount of work required.

Procedures for algal food culture have also undergone a series of changes. For several years, rather than relying upon 'wild' cultures

Fig. 5.4 Hatchery facility at Island Pond.

fertilized with pond bottom water, larval food needs were satisfied by fertilizing the outdoor tanks with the commercial fertilizer Miracle-Gro and inoculating the tanks with the Tahitian strain of *Isochrysis galbana,* a small flagellate, obtained from regional laboratories. *Isochrysis* was usually the dominant species in the culture, but it was supplemented by small diatoms and flagellates endemic to the pond. While this system gave reasonably satisfactory results for several seasons, the cultures in the outdoor tanks were sensitive to weather conditions and therefore inconsistent. Furthermore, it became apparent that oysters spawned relatively early in the spring, e.g. late April or early May, almost invariably did better than those groups spawned later in the spring or early summer, for reasons that remain unclear. In order to initiate spawning earlier in the season, it was necessary to have a relatively large and reliable supply of algal food available for conditioning the adults as well as feeding the larvae.

In recent years, therefore, and with the invaluable help of Steve Malinowski, a local oyster grower and client of Ocean Pond Corporation, the hatchery structure and production system have been modified. The floor space has been enlarged (from 8 to 15 m^2) and the walls insulated (Fig. 5.4). Algal cultures are initiated indoors with stock inoculants, progressing from small 1 ℓ flasks to 20 ℓ carboys to 80 ℓ fiberglass tubes under artificial light (Fig. 5.5). Under normal conditions, one tube is more than sufficient to satisfy the daily food requirements of the larvae and post-larval stages. Although adults being conditioned and early-stage larvae are fed only *Isochrysis*, a higher

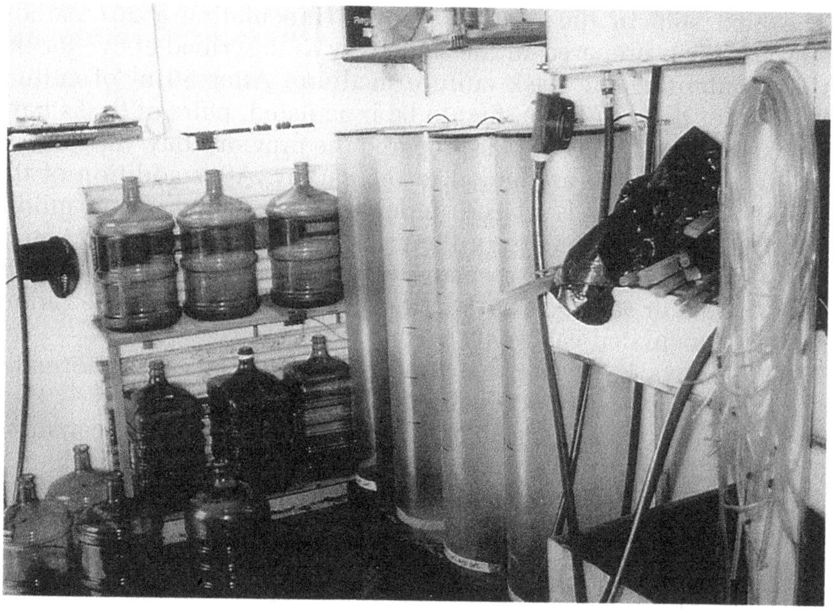

Fig. 5.5 Algal culture containers in hatchery.

rate of survival through setting has generally been obtained in our experience when the diet of the mid- to late-stage larvae includes a small diatom, such as *Chaetoceros neogracile* or *C. muelleri*, in addition to *Isochrysis*.

The procedure for algal culture is a version of the batch system and proceeds as follows. A 20 ℓ carboy is filled with pond water that has been passed through a 1 μm cartridge filter. The water is sterilized by adding 9 ml of sodium hypochlorite and the carboy vigorously shaken. After 1 hour, the water is dechlorinated by the addition of 3 ml of normal (1 N) sodium thiosulfate solution[5.1], and enriched with 6 ml of pre-prepared nutrient medium (Fritz F/2). This provides the basic media for the flask cultures.

Two flasks are each filled with 450 ml of the basic medium and autoclaved. Each flask is then inoculated with 50 ml of a Tahitian *Isochrysis* stock inoculant, usually obtained from the Cornell University Shellfish Laboratory on Long Island, New York, and held under fluorescent light at room temperature. After approximately 5 days, the cell concentration in the two cultures has increased to about 2 000 000–4 000 000 cells per ml, at which point 50 ml of culture is drawn from each flask and used to inoculate another pair of flasks prepared in the same fashion. This procedure is repeated each day until there are five pairs of flasks under light and an additional pair that has been washed, sterilized and prepared for inoculation the following day.

The next step in the routine involves inoculating a 20ℓ carboy, which has been prepared in the same way as described above for the carboy containing the flask culture medium. After 50ml of culture from each of the oldest, i.e. first to be inoculated, pairs of flasks have been used to inoculate the pair prepared the previous day, the remaining 450ml from each are added to the carboy. After addition of the nutrients, the carboy is placed under fluorescent lights and mildly aerated. As the cell density in the carboys increases, the aeration is increased as well in order to ensure sufficient exposure of the cells to light. With six or seven carboys in the culture system, a single carboy is usually inoculated each day.

After 5–6 days, the algal concentration in the oldest carboy normally is up to 2000000–4000000 cells per ml. The entire carboy is then used to inoculate one of the 80ℓ fiberglass tubes, which has been filled with 60ℓ of filtered pond water, sterilized with sodium hypochlorite and dechlorinated with sodium thiosulfate, and enriched with nutrient media, all at triple the concentrations described above for the carboys. After 4–5 days, the algal concentration in the oldest tube should be up to 2000000–4000000 cells per ml and ready for harvesting.

The culture procedure for *C. muelleri* or *C. gracile* proceeds in the same fashion, except that sodium silicate must be added to the culture medium. A standard stock solution is prepared by adding 15g of sodium silicate to 1ℓ of distilled water. To every 1ℓ of culture medium is then added 1ml of stock solution.

Air temperature in the hatchery is maintained, as closely as possible, between 20 and 25°C. Unfortunately, this is not always possible, and, particularly during extremely hot weather, the cultures in the carboys and tubes may crash, and the larvae and/or juveniles must survive on less food for a few days until the system recovers. Under normal conditions, the system proceeds with little variation and generally requires no more than 1 hour for one person to maintain.

The process for conditioning adult oysters for spawning begins in mid-April, and the first spawning is induced in early May. The broodstock consists of a strain of oysters developed for resistance to the disease MSX (*H. nelsoni*) by Rutgers University[5.2]. In 1985, a small sample of these oysters was obtained from David Relyea, manager of the Frank M. Flower oyster company. These were members of the sixth generation of this particular resistant strain, which had originated from Long Island Sound and been bred for resistance in the Delaware River. Each spring, a small group of the fastest growers from the previous spring's spawning, i.e. when 1 year old, are selected and retained as parent stock for the following spring. Approximately 50 of these oysters are brought into the hatchery in early April and placed in a long, shallow tank of recirculating pond water. By means of electric immersion heaters, the water temperature in the tank is elevated from ambient pond temperature, usually about 10°C this time of year, to about 18°C, at which it is held for 10–14 days. During this period, the oysters are fed *Isochrysis* at a rate of about 10^9 cells per oyster per day, and the tank water is changed daily. When the gonads appear to be fully ripe, the oysters are induced to spawn by slowly elevating the tank water temperature to 29°C, and adding a small suspension of sperm and/or eggs obtained from one of the oysters. After spawning, the oysters are put back into the pond. Often the same group may be used for a second spawning.

As soon as an oyster begins to spawn, it is removed from the tank, rinsed in clean water, and allowed to continue spawning in a separate container. When what appears to be a sufficient number of eggs 30 000 000 or so, have been discharged, these are poured into a larger container and mixed with small quantities of sperm from all the males. The purpose of this is to attempt to diversify the genetic make-up of the eventual offspring and thereby reduce the chances of a high percentage of poor-quality oysters.

Within 1 hour of fertilization, the fertilized eggs are transferred to two of the four 400 ℓ conical tanks. In preparation, the pond water in the tanks has been passed through a 10 μm bag filter and pre-heated to about 25°C. The desired density at this stage is about 25 fertilized eggs per ml, or about 10 million per tank. The water is mildly aerated in order to prevent the embryos from settling to the bottom. From this point on, the water in the tanks is changed, the tanks are scrubbed, and food is added daily. The tank water being discarded is passed through a fine mesh sieve, initially of 44 μm porosity, in order to collect the larvae, which are then transferred to a tank of clean water. During the spring, when water temperatures are low and the water needs to be pre-heated, the two pairs of tanks are used on a rotational basis, one for holding the larvae, the other for heating water. Throughout the larval period, the tank temperatures are maintained at about 25°C.

Food is added to the larval tanks at a rate of about 10–20 ℓ per tank per day, depending upon the density of the algal cultures. The objective is to provide a concentration of about 10^5 algal cells per ml in the tanks. If, after 24 hours, the water in the tanks is clear, slightly more food is added; conversely, if the water is not clear, the amount of food is reduced. Generally speaking, over-feeding is to be avoided as much as under-feeding, since it can seriously inhibit larval growth. Diatoms are not included in the main diet of *Isochrysis* until the larvae begin to exceed 200 μm in size.

As the larvae grow, sieves of increasing mesh size are used when the tanks are drained for cleaning, the objective being to get rid of stunted larvae that will not grow appreciably but which occupy space and take up food. As a result, the numbers of larvae in the tanks gradually decrease, from an original 10 000 000 to 2 000 000–4 000 000 per tank by the time they have reached a setting size of about 275–300 μm. This growth period, between the time the eggs are fertilized to when the first larvae reach setting size, may vary from 8 to 12 days. Larvae that are retained on a 210 μm screen are transferred to the setting tanks. Those that pass through are returned to the larval tanks for further growth.

The handling of larvae at metamorphosis is perhaps the most critical time in the hatchery operation. In order to assure a high percentage of single oysters, it is necessary to use a cultch material finer than the dolomite or shell particles that were once used and were large enough to be retained on fly screen; therefore oysters that have recently set on the so-called 'micro-cultch' must be held in containers with a mesh much finer than fly screen.

Two procedures have been used at Island Pond: placing the ready-to-set larvae in fine mesh floating trays under static conditions, or placing the larvae in a downwelling system. The static system simply involves positioning a series of shallow 20 × 20 cm trays with 202 μm mesh bottoms in a rectangular tank filled with pre-heated pond water. Finely ground oyster shell is spread thinly over the bottom of the setting trays, the larvae are placed in the trays at a density of about 200 000 per tray, and the water temperature is maintained at 29–30°C. The water in the main tank is drained every day, the tank is hosed down and refilled with fresh pre-heated pond water, and algal food is added. The mixed diet of *Isochrysis* supplemented with *Chaetoceros* is provided in sufficient amounts to provide a minimum concentration of 10^5 cells per ml. The newly set oysters are held in this system for 7–10 days, by which time they are usually large enough to be moved outdoors to the upwelling system.

The alternative procedure is to transfer setting-size larvae into a downwelling system, also employed at many hatcheries. This consists of a series of PVC cylinders, or silos, approximately 30 cm wide and 30 cm deep, with fine mesh (202 μm) bottoms, which are suspended in indoor tanks filled with pre-heated pond water. Cultch particles are spread in a thin layer on the bottom of the silos before the larvae are introduced, at a density of about 200 000 per silo. Water is continuously circulated down through the cylinders by means of an air-lift system linked to the hatchery air supply, providing the larvae and post-larvae with a constant supply of food. As in the static system, algal food is added daily.

When setting is completed and many of the juvenile oysters have reached 500 μm or so in size, they are transferred to silos held in outdoor tanks into which pond water is pumped continuously. In this type of once-through system, the water flows upwards through the bottoms of the silos and is then returned to the pond[5.3]. In the upwelling system at Island Pond, the tanks measure 0.6 m × 0.6 m × 2.4 m long, and water enters the tanks at a rate of about 80 ℓ per minute. The initial stocking density does not exceed 250 ml of juveniles per silo. At this stage, no supplemental food is added, and the juveniles subsist upon whatever planktonic food is available in the pond water.

Twice each week, the oysters are sieved through a series of screens – 1.0, 0.75 and 0.5 mm mesh – and returned to silos having the appropriate mesh size. Those retained on the 1.0 mm mesh screens are transferred to the floating fly screen trays and placed in the pond.

Using cultch large enough to be held in fly screen trays, as was done in earlier years, requires less handling and maintenance during the early juvenile stage than the 'micro-cultch' system described above, involving the operation of an upwelling system. The upwelling tanks must be emptied and cleaned once each week, and the silos containing the juvenile oysters are cleaned every other day. However, the additional time is justified by the improvement in quality of the seed, i.e. absence of clusters.

Within 2 weeks of spawning, setting usually has been completed, and, after 3–4 weeks, all of the juveniles have been transferred from the hatchery to the outdoor upwelling system, allowing room in the hatchery for a second batch of larvae. The amount of time that the juveniles remain in the hatchery depends upon water temperature in the pond; the lower the temperature, the longer they are retained indoors. This process is repeated once more, providing a total of three separate batches of larvae by the end of June.

Regardless of the relative numbers of larvae reared at different times during the May–June period, highest survival nearly always occurs among the earliest batches. Percentage survival of larvae between the time they reach setting size and the time they are large enough to be transferred to fly screen trays rarely exceeds 25–30% and is usually lower. Many are lost in the setting process by attaching to the surfaces of the containers, and there may be additional losses to predation in the upwelling system, such as by the flatworm *Stylochus*.

Since the facility is not well equipped for hatchery operations prior to April, broodstock have often been sent to other hatcheries in late winter, where they are spawned and the larvae reared through setting. By the time this batch is returned, the pond water is usually warm enough, 12°C or above, for them to be put directly into the outdoor upwelling system, but there is little sustained growth until temperatures reach 15°C.

During the mid-1980s, the advantages of producing and selling oysters that had been grown in the pond for two growing seasons, rather than just one, were explored. The primary objective was to determine the technical and economic feasibility of suspension culture on a commercial scale. This project was funded in part by the New York State Urban Development Corporation, which at that time was interested in the economic potential of aquaculture development in the state. Off-bottom culture of shellfish assures a higher rate of survival than the traditional on-bottom methods, which usually incur heavy losses from predation and/or siltation. Furthermore, appropriation of natural shellfish areas for private use is generally opposed in New York State by shellfishermen, particularly where this might involve grounds naturally favorable for shellfish. Since the nature of the bottom is generally not important in off-bottom culture, this method offers greater flexibility in terms of selection of culture sites and reduces the likelihood of conflicts.

For purposes of culturing the oysters in the pond for a second season, five-tier lantern nets with 12 mm mesh were selected, and yearling oysters obtained from a spawning of wild pond oysters were used. For larger oysters, growth rate in these nets compared favorably with that in the surface trays, and a larger number of oysters could be contained in each lantern net than in an equivalent number of pearl nets. Growth comparisons of oysters held at different densities, at an initial size of slightly over 1 inch, indicated that a loading density of 200 oysters per tier (1000 oysters per net) produced the best results in terms of ultimate yield.

By the end of the growing season, these oysters had roughly doubled in size with respect to growth of shell, with very little mortality. However, none had reached a minimum acceptable marketable size of 75 mm. Furthermore, the tendency of many of the oysters to grow into the meshes of the net, combined with fouling of the nets by algae and other marine organisms, including wild oysters, required continual maintenance and added significantly to the overall cost of the operation (Fig. 5.6). It was also found, by the end of the growing season, that many of the oysters were heavily infected by *Polydora*, the polychaete worm which causes unsightly blisters on the inside of the

Fig. 5.6 Lantern nets being air-dried.

oyster's shell. Of interest, however, was the fact that oysters from the same original group that had simultaneously been cultured in lantern nets, and at the same density, in a high salinity area with appreciable tidal currents, e.g. Fishers Island Sound, did significantly better with respect to shell growth. Nearly 20% of the oysters reached minimum market size during that same period. A successful off-bottom oyster culture operation, utilizing lantern nets, has since been established at this same site by Steve Malinowski.

The feasibility of culturing oysters in the pond for a third growing season was also explored, using five-tiered lantern nets with 25 mm mesh. Approximately 80% of these oysters, grown at a density of 100 per tier, reached marketable size by the end of the growing season, when 30 months of age. Although the pond oysters had a more favorable meat weight : total weight ratio than their siblings grown in Fishers Island Sound, the Sound oysters had developed a harder, cleaner shell, and the incidence of *Polydora* blisters was much lower.

In 1985, Ocean Pond Corporation abandoned the use of wild pond oysters as broodstock and has since concentrated upon breeding the Rutgers University strain described above. This particular strain of oysters has proven to be distinctly superior to the native pond oysters in terms not only of disease resistance but also of growth rate. A significant percentage of each year's production may reach marketable size when only 18 months of age. However, because of periodically high coliform bacteria counts in the pond, attributed to an abundant waterfowl population, Island Pond cannot be certified for purposes of harvesting oysters for direct human consumption. Therefore its best use is as a source of seed oysters, with good growth characteristics and disease resistance, that can be provided to growers with certified grounds and higher salinity water, where the oysters can develop a harder, cleaner shell and saltier flavor.

6 Oyster Culture in the Indo-Pacific Region

In comparison with Japan, Korea and People's Republic of China, oyster production in the Indo-Pacific region is relatively small. In the FAO fishery statistics for 1995, only six countries in this region are listed as oyster producers: Indonesia, Malaysia, Philippines, Thailand, Australia and New Zealand. Combined total landings from these countries in 1995 amounted to about 53 000 t, or less than 2% of the world total. Nevertheless, in view of the vast amount of undeveloped shore line in the south-western Pacific region, there would seem to be great potential for increased production in the future.

The history of oyster culture in Australia has been well documented[6.1]. The industry today is based upon three species: the Sydney rock oyster (*Saccostrea commercialis*), the Pacific oyster (*C. gigas*) and the native, or mud, oyster (*Ostrea angasi*). Of the three, the Sydney rock oyster is still the most important, presently accounting for more than half of the total oyster harvest, but there has been relatively little increase in production of this species over the past twenty years (Fig. 6.1). Annual production of the Pacific oyster, on the other hand, has increased markedly, i.e. by a factor of five, during the past decade (Fig. 6.2). The natural populations of *O. angasi* were exploited heavily during the nineteenth century, and the surviving populations are very small and contribute little to total production. Two additional species that are farmed in Australia are the milky oyster (*Saccostrea amasa*) and blacklip oyster (*Saccostrea echinata*), both native to Queensland[6.2]; as in the case of *O. angasi*, their annual harvest is very small.

The rock oyster is native to the east coast of Australia, its range extending from Queensland to Victoria[6.1]. The major producing areas traditionally have been Moreton Bay in southern Queensland and the rivers and estuaries of New South Wales, primarily the Georges River, Hawkesbury River, Port Stephens and Wallis Lake. The habitat of this species is primarily intertidal, partially because subtidal populations have been eliminated by predators and parasites. It is not particularly large, mature individuals rarely exceeding 75 mm in size. It is likely that its comparatively small size is partially a reflection of its intertidal habitat and consequent abbreviated feeding period. The two valves are markedly different, the left (lower) being deeply cupped and fluted, the right small and flattened. These oysters generally reach marketable size in 3–4 years.

This is a warm-water species that usually spawns when water temperatures reach 25°C, and spawning may continue intermittently for

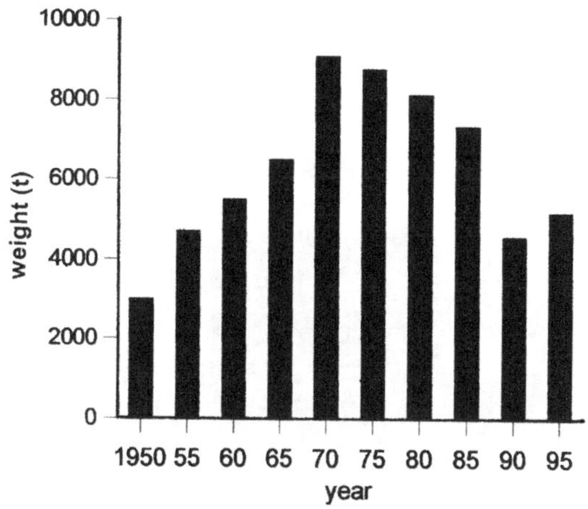

Fig. 6.1 Oyster production (whole weight) in Australia.

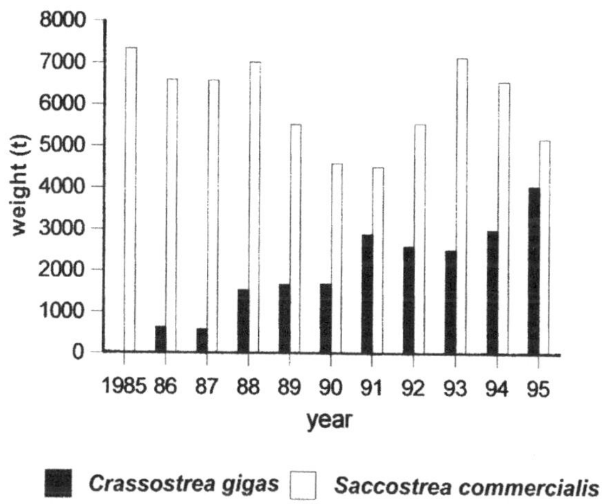

Fig. 6.2 Trends in production of Sydney rock oysters (*Saccostrea commercialis*) and Pacific oysters (*Crassostrea gigas*) in Australia.

several months. Spawning is often triggered by a sudden rise in water temperature, an abrupt drop in salinity, or a combination of the two. Although this oyster prefers relatively high salinities, in the 25–35‰ range, the larvae are quite tolerant of low salinities and may settle on a suitable substrate – rocks, shells, mangrove roots, etc. – considerable distances upstream from the river mouths.

The initial methods of culturing the rock oyster were quite primitive[6.2]. Because oysters that had set below the low tide mark suffered extremely high mortalities, a variety of cultch materials – stones, galvanized wire on wooden frames, and shells – were initially set out in

Fig. 6.3 Wooden stakes used for collecting spat, Australia (photo by I. Cameron).

intertidal areas. Natural setting was often supplemented by knocking oysters off the roots of mangrove trees and placing these on the shell beds or galvanized wire. During the 1930s, the growers began to use tarred wooden sticks, made from *Eucalyptus* trees, as cultch, and this method has persisted to today. The sticks, about 2 m in length, are nailed onto wooden frames (Fig. 6.3), and the entire frame is then dipped into tar and allowed to dry for several weeks. The purpose of the tar is to protect the sticks from damage by shipworms (*Teredo* spp.). A large producer may put out over 500 000 sticks each year.

The major spat-collecting area traditionally has been Port Stephens (Fig. 6.4), about 130 miles (209 km) north of Sydney, where about 80% of the rock oysters produced in Australia usually originated. In earlier times, favorable sets could be obtained in most of the river systems of New South Wales. However, the combination of siltation and biofouling in most of these estuaries forced the growers to depend upon the relatively clear and clean waters of Port Stephens for annual recruitment. It became evident in the mid-1980s that the Pacific oyster (*C. gigas*) had become established in Port Stephens, either through accident or design, and, as a result, certain restrictions were placed on moving seed out of this estuary in order to prevent the spread of *C. gigas* to other estuaries in New South Wales, the fear being that it would crowd out the slower-growing *S. commercialis*.

The oyster growers of New South Wales lease ground at Port Stephens for purposes of collecting their seed. On their individual

Fig. 6.4 Spat-collecting racks, Port Stevens, Australia (photo by O. H. Matthiessen).

plots, long racks about 1 m wide and running in parallel lines from the foreshore into deeper water, are assembled to support the frames, or lattices, of tarred sticks above the low water mark. The frames are stacked in tiers on the racks and are wired to the tops of the racks to prevent dislodgement by waves or rising tides. By positioning the sticks intertidally, fouling by barnacles and mussels is much reduced, and predation on the spat by finfish is less likely.

The collectors are left on the setting racks for varying lengths of time, depending upon the growth of the spat and the location of the growing areas. Some growers may begin to transfer the sticks to their own grounds when the oysters are 2–3 months old and roughly 12 mm in size. Near the site of the growing leases, the collectors are again placed on racks high in the intertidal zone, the lattices again stacked in tiers to discourage predation by finfish. In these so-called 'depot areas', the stacking of the frames of sticks in tiers not only discourages finfish predators but also protects many of the spat from direct exposure to the sun.

After about 6 months in the depot areas, the frames of sticks are transferred to the growing racks positioned lower in the intertidal zone. The young spat, hardened by relatively long exposure to air during depoting, are now more resistant to predation. Growth is encouraged by arranging the frames singly rather than in tiers in order to maximize the flow of current around each stick. The oysters remain

Fig. 6.5 Mature oysters on growing racks, Australia (photo by I. Cameron).

on these growing racks for a further 2 years, at which time many will be ready for harvest (Fig. 6.5). The position of these sticks with respect to the low water mark may be critical. If too low in the intertidal zone, fouling may be excessive. If too high, growth rate is retarded, and exposure to the sun may be lethal. This species cannot tolerate exposure to air at temperatures in excess of 35°C for prolonged periods, and adult oysters appear to be more vulnerable to high temperature than the juveniles.

Although the Sydney rock oyster is harvested and sold, in the shell, to dealers year-around, the primary oyster season is in early summer, before the oysters have spawned and when the gonads are well-developed. During harvesting, the oysters are knocked off the sticks on shore and culled. The larger oysters are sold to processors, who open the oysters and either sell them in the shell to restaurants or, in the case of the smaller oysters, shuck them and sell the meat in bottles. Oysters knocked off the sticks that are too small for processing are either sold to other growers to raise to a larger size or kept by the initial grower for the same purpose.

Further growth of these small oysters, referred to as 'seconds', takes place in trays supported off the bottom on the intertidal racks. These are open-top trays constructed of wooden frames and wire mesh dipped in tar. In order to reduce oyster mortalities resulting from heat during summer, the trays are often covered by plastic mesh and may be sprayed with sea water during exceptionally hot weather (Fig. 6.6).

Fig. 6.6 Spraying during warm weather to cool and clean oysters, Australia (photo by O. H. Matthiessen).

The oysters in the trays are also sprayed periodically to remove silt that accumulates in the trays and which encourages the incidence of *Polydora*, a marine worm that invades the oyster's shell, causes unsightly blisters on the inner shell surface, and may eventually kill the oyster by weakening the adductor muscle.

Because of incidence of oyster-related gastrointestinal disease, all oysters sold in New South Wales or shipped overseas must first be depurated. The depuration process consists of holding the oysters for a period of time in tanks filled with sea water that has been sterilized by ultraviolet light. The holding period in the tanks is 48 hours. Samples of oysters are taken from the tanks after treatment at periodic intervals by government inspectors and analyzed for coliform bacteria in order to ensure that the oysters are being properly cleansed.

It is evident that losses during the 3–4 year growing period can be severe. A major problem in New South Wales is a disease known as 'winter mortality'[6.3], caused by a protozoan, *Mikrocytos roughleyi*. This parasite appears to be linked with lower-than-normal salinities, since major mortalities occur after periods of heavy rainfall[6.4]. In some years, up to 80% of the marketable oysters in certain growing areas may die. Heavy mortalities have also been reported among oysters in Moreton Bay in Queensland, where the cause of death has also been attributed to a parasitic protozoan, *Marteilia sydneyi*. Significant oyster predators include certain species of fish and crabs. Finally, substantial losses may occur during very warm weather since the oysters are exposed to the air at low tide.

The scarcity of suitable intertidal growing areas, the losses that occur because of exposure during hot weather, the labor involved in installing and maintaining racks, and the relatively long time required to reach marketable size on the intertidal racks have combined to stimulate interest in deep-water culture of this species in recent years. A variety of methods have been tested, including the Japanese system of hanging wire strings, shallow trays hung from surface pontoons, tiers of trays suspended beneath rafts, and wire cages that may be slung from rafts or suspended from surface floats[6.5,6.6].

Each of these systems has been found to have drawbacks. The labor costs involved in assembling the shell strings, or 'ren', were considered by many oyster growers to be prohibitive. The mortality rate among oysters held in trays, singly or in tiers, beneath pontoons or rafts was found to be excessive in some areas. In certain cases, losses were attributed to 'winter mortality'. At other locations, fouling by barnacles, tunicates, sponges, hydroids and other organisms became a serious problem. Also, the incidence of *Polydora* was frequently found to be too high.

Despite these disadvantages, a small number of oystermen in New South Wales have abandoned the traditional method of stick culture in intertidal areas in favor of off-bottom, deep-water culture techniques. Subtidal culture in some areas at least has been shown to accelerate growth rate, improve meat quality, and avoid excessive heat conditions[6.2]. Although certain of the deep-water growers appear to be successful, it seems clear that success or failure is site specific to a large extent. In some areas, the incidence of *Polydora* is so heavy as to preclude subtidal culture, and the dangers of excessive currents during floods, heavy siltation and fouling, predation and disease must all be considered. There is hope, however, that hatchery production of triploid oysters, oysters possessing an extra complement of chromosomes, may eventually reverse the industry's downward trend, since these oysters appear to grow more rapidly and have greater resistance to disease than wild diploid oysters[6.7].

The outlook for the Pacific oyster industry, on the other hand, appears to be more encouraging. This species was deliberately introduced into Tasmania from Japan in 1947 in an effort to revive an industry ruined by excessive exploitation of the native oyster (*O. angasi*) beds[6.8]. After a series of plantings in northern Tasmania, natural sets began to occur[6.9], and eventually its distribution extended into the Tamar River, where it began to replace the remaining native oysters and become the basis for a new oyster industry.

Initially, the method of culture of the Pacific oyster in Tasmania was similar to that used in the culture of the Sydney rock oyster in New South Wales. Wooden sticks were tarred, assembled on frames, and placed upon intertidal racks during the spawning season, which usually began in early January. After several months, the sticks with spat attached were transferred to the growing racks, which were positioned intertidally in favorable growing areas. It was found that, because the growth rate of this species is extremely rapid and because predators are relatively scarce in Tasmania, depoting of the juveniles, as practiced in New South Wales for the rock oyster, was not necessary.

Within 16–30 months, the oysters reach a size of about 75 mm and are ready for market[6.8]. The oysters are knocked off the sticks, graded for size, and the larger ones are sold to processors. Smaller oysters dislodged from the sticks in the process are placed in intertidal trays for further growth.

Certain of the oyster growers have used strings of scallop (*Pecten meridionalis*) shell to obtain spat and to grow the oysters to market size, in very much the same fashion as the Japanese. In order to collect the spat, the strings of shell are suspended from rafts to a depth of 2 m. Approximately 3–4 months after settlement, the shell with spat attached is restrung on wire or synthetic rope, with spacers – 20 cm lengths of garden hose – separating the shells. The strings are then hung from rafts or long-lines, and, because the oysters are constantly submerged, marketable size is obtained in 12–24 months.

More recently, because natural reproduction in the Tamar River was not sufficiently reliable, the Pacific oyster industry in Tasmania has come to depend heavily upon seed production in hatcheries. Emphasis is upon the production of single oysters, using tiny shell fragments as cultch. Usually, the grower receives his seed from the hatchery when only 2–3 mm in size and places these in plastic mesh bags, envelopes or baskets. These in turn are placed on, or suspended from, intertidal racks (Fig. 6.7). As the oysters grow, they are shifted, at lower densities, to bags of coarser mesh. This system is very similar to, and was adapted from, the French method of intertidal culture. Predation is rare, and survival is generally high. The oysters are ready for market within approximately 18–24 months, which is considerably faster than the 36 months or more required for the Sydney rock oyster.

According to a report published in 1993[6.10], the hatchery of a large oyster enterprise in Tasmania was producing more than 40 000 000 *C. gigas* seed each year, providing stock for the company's own leases as well as for other oyster growers in Tasmania and South Australia. This volume could readily double should the need arise. After setting, the oysters, 0.7–1 mm in size, are transferred to a series of upwellers and grown to a size of 3–4 mm. Small ponds adjacent to the hatchery facility are fertilized, creating the growth of microalgae that are pumped through the upwelling system as food. The oysters are then moved to outdoor floating upwelling systems, where they grow to 10–15 mm. They are then transferred to the company's intertidal leases and placed in plastic mesh bags supported off the bottom on racks. When the

Fig. 6.7 Plastic mesh bags used for growing single oysters off-bottom (photo by I. Cameron).

oysters reach a size of 45–50 mm, they are transferred once again, this time to trays which are suspended from long-lines in deep water. The oysters are brought ashore every 6–8 weeks, where they are cleaned and graded for size, and then returned to the offshore trays to complete growth to marketable size.

Until 1990, the Pacific oyster was regarded as a pest by the New South Wales oystermen, and its introduction into this state was not allowed. However, introductions, accidental or otherwise, did occur, and eradication measures have not been successful[6.11]. At the present time, its culture is allowed in certain areas of New South Wales. It is also cultured successfully in South Australia and West Australia, where the industry relies largely upon hatchery production.

There are three species of oysters harvested in New Zealand: the Auckland rock oyster (*Saccostrea glomerata*), the Pacific oyster (*C. gigas*), and the New Zealand dredge oyster (formerly *Ostrea lutaria* and now placed under a separate genus and designated *Tiostrea chilensis*). Until recently, the most important of the three species from an economic standpoint has been the dredge oyster, which is not cultured but simply harvested from wild stocks in relatively deep water. This fishery, at the southern tip of New Zealand, has now declined, and the relatively small New Zealand oyster industry is based largely upon the ubiquitous Pacific oyster. The traditional oyster of culture in New Zealand is the native rock oyster, which is very similar to the Sydney rock oyster in appearance and natural history.

The Pacific oyster was first identified in New Zealand in 1971[6.12]. This species had apparently been introduced inadvertently by way of the bottoms of cargo ships from Japan. Because the water temperatures of northern New Zealand are favorable for the reproduction and growth of this species, it became well established within a very few years and, in many areas, crowded out the native rock oyster. At the present time, the great majority of oysters cultured in New Zealand are of this species.

The culture of oysters by private growers did not get under way in New Zealand until 1964, when the policy of issuing private leases was first initiated[6.13]. The original procedure was to collect wild oysters from wherever they had settled naturally, and hold these in bottom trays postitioned in the intertidal zone until they reached market size. This method was soon superseded by the intertidal stick method, adopted from Australia, which is used at the present day. Various types of stick materials have been tested in New Zealand, including asbestos, cement, reinforced concrete, aluminum, tarred hardwood, and tarred hardwood that has been dipped in cement.

The oysters are considered to be in prime condition for harvesting immediately prior to spawning. As in Australia, marketable rock oysters are 3–4 years of age, while the Pacific oyster can reach harvestable size when only 18–24 months of age[6.14]. Oysters knocked off the sticks that are too small for market are transferred to trays, supported on racks in the intertidal zone, for further growth. In the case of the Pacific oyster, it has been found practical to transfer small oysters knocked off the sticks during harvesting to plastic mesh bags rather than exposed trays. These are placed either on intertidal racks or on the bottom in the intertidal zone and turned over periodically.

In the Philippines, shellfish have been farmed for nearly a century[6.15]. During the past decade, annual production has ranged between 10 000 and 20 000 t (Fig. 6.8). The average oyster farm is very small, ranging from less than 1 ha to 4 ha in size. Several species of oysters are harvested for food in this region, the most important being the slipper oyster (*Crassostrea iredalei*) and, to a lesser extent, a smaller species, *Crassostrea malabonensis*. Favored by warm water temperatures year-around, the slipper oyster may reach a size of over 75 mm in 6–9 months.

Basically four different types of culture are used in the Philippines, broadcasting on the bottom, stake culture, lattice culture and hanging culture[6.16]. The broadcast method, which requires the least amount of time and materials, is used only in areas where the bottom is firm enough to support spat collectors, which may consist of stones, tin cans, logs or oyster shells. The collectors are spread on the bottom in areas known to have good oyster sets. Once the cultch material has received a set, it is moved to areas favoring more rapid growth. Usually the oysters

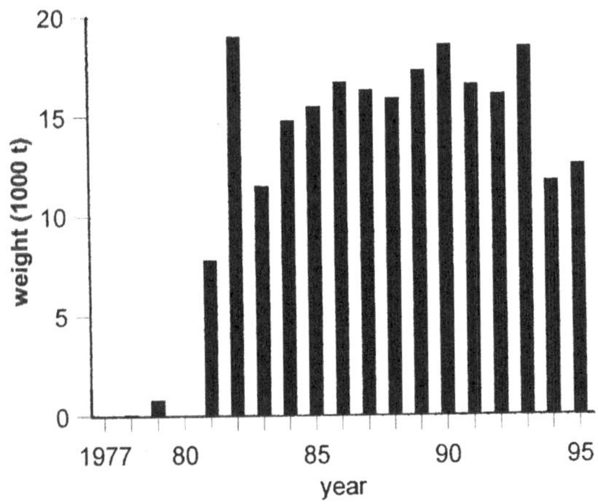

Fig. 6.8 Oyster production (whole weight) in the Philippines.

are large enough to be harvested within 8–12 months. Although this method requires a minimum amount of capital outlay and maintenance, mortality from bottom predators may be high and harvesting less efficient than would be the case for off-bottom culture[6.17].

In areas where the bottom is soft, such as Manila Bay, the stake method is more often used. The stakes, usually of bamboo, are driven into the bottom in rows, spaced about 0.5 m apart, at the beginning of the spawning season. In order to increase the surface area for setting, oyster shells may be hung from the stakes, or horizontal strips of bamboo are fastened to the tops. Although the yield by this method is higher than that achieved by the broadcasting technique, owing to less mortality caused by bottom-crawling predators, bamboo does not provide an ideal surface for attachment of spat. Furthermore, crabs, starfish and boring snails eventually climb up the stakes to feed on the oysters.

A third method involves weaving lengths of split bamboo into lattices. The lattices can be positioned either horizontally or vertically in intertidal areas, erected as tents or in fences, or suspended in the water from racks. The lattices are readily moved, with juvenile oysters attached, to favorable growing areas after setting.

The hanging method generally utilizes oyster shell, coconut shell or strips of rubber tires as cultch. These materials are strung on monofilament line, with knots or short lengths of bamboo as spacers to separate the cultch. The strings are hung from horizontal posts supported by uprights, from lines strung horizontally between posts, or from rafts. This method helps minimize predation and maximize growth rate and quality of meat. However, the materials may be too costly for some oyster farmers.

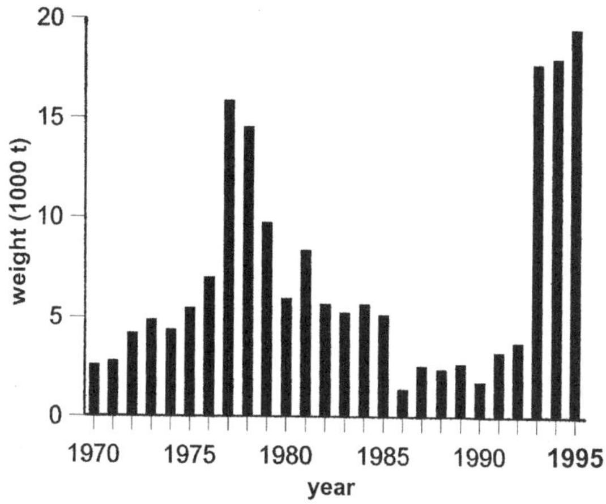

Fig. 6.9 Oyster production (whole weight) in Thailand.

Although these methods are primitive, they are inexpensive, and production is second only to Thailand in the Indo-Pacific region. It has been projected that, with such an extensive shore line, production in the Philippines could be greatly increased[6.15].

In recent years, the major oyster producer in the Indo-Pacific region has been Thailand, with annual landings of about 20 000 t tons (Fig. 6.9). Production has, however, varied widely during the past 30 years. There are three species of commercial importance: *S. commercialis*, a small oyster, and two larger species, *Saccostrea lugubris* and *Crassostrea belcheri*. All three species occur in varying concentrations along Thailand's 2500 km coastline[6.18].

Oyster culture in Thailand was apparently initiated about 50 years ago by immigrant Chinese who began to place stones on the bottom as cultch, and this method is still pursued[6.18]. The rocks are piled in rows in intertidal areas. In areas of soft bottom, the rocks are placed on mats of bamboo to prevent sinking. Other cultch materials include cement blocks, car tires, asbestos roof tiles, bricks and pottery. In some areas, cement poles are used separately or in conjunction with rocks. These poles, about 0.5 m in length, with a hole at the lower end, are slid down over wooden poles imbedded in the bottom sediment. In other areas, cement pipes may be used as well as bamboo stakes.

Off-bottom culture consists primarily of the use of trays mounted on stakes. This system is used for the most part for oysters too small for market that have been removed from rocks or other cultch materials during the harvesting process. Although the Japanese method of suspension culture, i.e. hanging strings of oysters from rafts, has been tested in Thailand, it has not been particularly successful owing to the cost of materials and the fact that much of Thailand's coast is exposed to wind and wave action. In view of the extensive area available for culture, however, production would be expected to increase.

Although some oyster production occurs in Indonesia and Malaysia, combined annual landings rarely exceed more than 1000–2000 t. Some experimental culture has been initiated, but most of the production derives from the harvest of wild stocks. In Malaysia, some success in the collection of wild *Ostrea folium* spat on polyethylene nets and ropes has been reported[6.19]. The oysters mature to a marketable size of 60–70 mm within 8–10 months.

7 Artificial Upwelling

An interesting feature of Island Pond is the nutrient-rich bottom layer of water that develops in the deeper areas of the pond during the spring and summer months. During the fall, when surface temperatures drop and the pond 'overturns', this layer disappears as the water column becomes nearly isothermal and isohaline. The months during which the pond is stratified are also those during which water temperatures are most favorable for oyster growth. However, this natural reservoir of nutrient material is essentially 'locked in' and unavailable for stimulating photosynthesis near the surface where the oysters are cultured.

Although much of the pond is stratified to some extent during the summer, the most pronounced physical and chemical gradient occurs near the south-east corner of the pond where, as might be expected, the water is deepest (Fig. 1.3). This part of the pond is near the opening leading into Beach Pond and Block Island Sound, and stratification results from periodic intrusion of relatively cold and saline sea water that sinks to the pond bottom because of its higher density. The fact that the temperature of the bottom layer during summer may actually be cooler than that of the Sound water might also suggest the influence of cold subterranean water entering the pond through the bottom sediment. The presence of springs in the pond is evident by openings or weak spots in the ice that persist during the winter.

Observations on this bottom layer were initiated in 1979 as part of an algal culture experiment to determine the feasibility of enhancing the growth of microalgae in outdoor tanks by enrichment with bottom water. This experiment was carried out jointly with the Woods Hole Oceanographic Institution under the National Sea Grant Program. During the period March–July, water samples were collected at a location in the pond where, on the basis of bottom soundings, depth was found to be greatest. The temperature and salinity of the samples were recorded, and they were analyzed for nutrient (phosphate, nitrate, nitrite, ammonia and silicate) and oxygen concentrations and pH.

As shown in Table 7.1, Island Pond was already strongly stratified with respect to temperature, salinity, and inorganic nutrient concentration (except nitrate) by mid-May, and nutrient concentrations at the bottom intensified during June and July. In an initial fertilization experiment, about 14 000 ℓ of pond surface water were pumped into an outdoor cylindrical tank, having a total volume of about 15 000 ℓ. The tank was partially submerged in the pond in order to maintain a

Table 7.1 Temperature, salinity, and nutrient concentrations (μm) with depth, May, 1979

			Nutrient concentrations (μm)			
Depth (m)	Temperature (°C)	Salinity (‰)	PO_4	Nitrate NO_3	NH_3	SiO_3
0	19.8	20.0	0.33	0.13	1.92	6.93
1	18.6	22.0	0.29	0.29	1.56	5.67
2	19.0	22.0	0.27	0.23	1.26	1.58
3	18.5	26.0	0.50	0.03	3.72	4.23
4	15.8	28.0	0.42	0.05	2.88	2.84
5	12.0	29.0	0.81	0.05	10.38	28.53
6	10.2	31.0	1.98	0.14	40.38	28.53

water temperature similar to that of the pond. Each day, between 640 and 960 ℓ of bottom water, pumped from the bottom of the pond into 400 ℓ containers, were brought to shore and added to the tank after an equivalent volume had been removed. These daily enrichments resulted in a significant increase in phytoplankton cell density within a matter of a few days (Fig. 7.1), far exceeding that of the surrounding pond water.

During these initial fertilization experiments in 1979, it was found that the algal cultures resulting from bottom water enrichment provided an excellent source of food for oyster larvae then being cultured in small containers at the pond site. At tank water temperatures below 24°C, the dominant algal species in the culture tank was *Thalassiosira* spp., a small centric diatom. A small microflagellate, similar to *Nannochloris* spp., predominated at higher temperatures.

It was evident from these preliminary investigations that nutrient concentration in the pond, at least with respect to phosphate, silicate and ammonia, was directly related to depth, with a major increase occurring about 6 m below the surface. This bottom layer of nutrient-rich water, localized in the deepest area of the pond, was therefore relatively thin, less than 1 m in thickness, with a significant nutrient gradient within the layer itself.

These results provided the basis for more extensive larva culture work during 1980 and 1981. During the larval season, i.e. June–July, algal food was provided by inoculating 400 ℓ outdoor tanks with bottom water at an initial ratio of 1:10 bottom water to surface water, the latter having been passed through a fine mesh bag filter in order to remove much of the zooplankton. The same volume of bottom water was added on the second and, if necessary, third days as well. This usually resulted in an algal bloom within 3–4 days, at which time the entire tank was harvested and a new culture started. Daily food requirements were satisfied by starting a new culture each day.

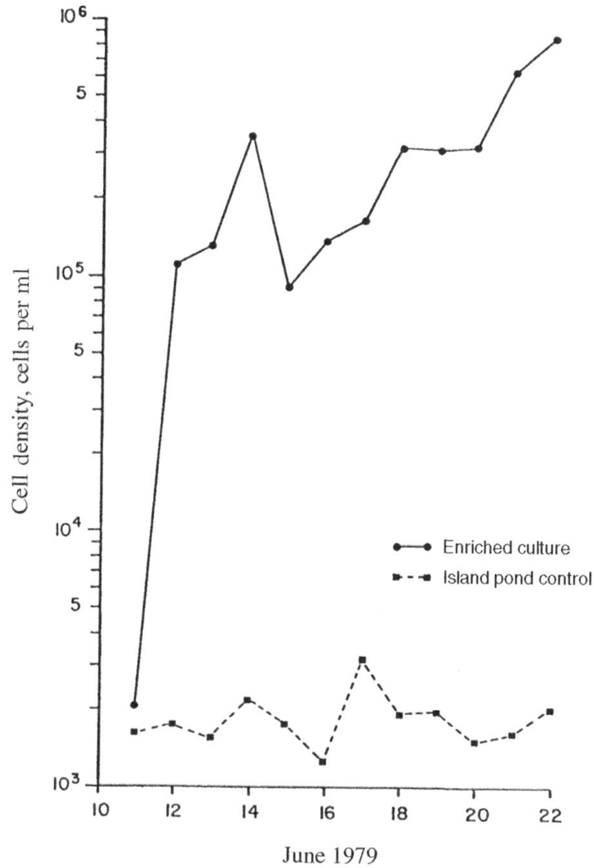

Fig. 7.1 Relative rates of growth of phytoplankton in fertilized and control tanks.

It was clear from these experiments that enrichment of surface water with bottom water had, at least on a relatively small scale, pronounced beneficial effects in terms of oyster food production. This in turn suggested the possibility that the entire pond might be successfully 'fertilized' in order to increase the production of algal food for oysters, not just for the larvae in tanks but also for the juvenile oysters grown in the pond. (More than half a century ago, a sea loch in Scotland was fertilized with inorganic nutrients in an attempt to increase the yield of flatfish and shellfish[7.1]. Although the productivity of the loch was initially increased, increased production could not be sustained without repeated and expensive applications of fertilizer, and the project was discontinued). The desirability of increasing the pond's natural food supply was suggested during 1980 and 1981. During 1980, an exceptionally large standing crop of phytoplankton developed in early spring and persisted throughout the summer and into

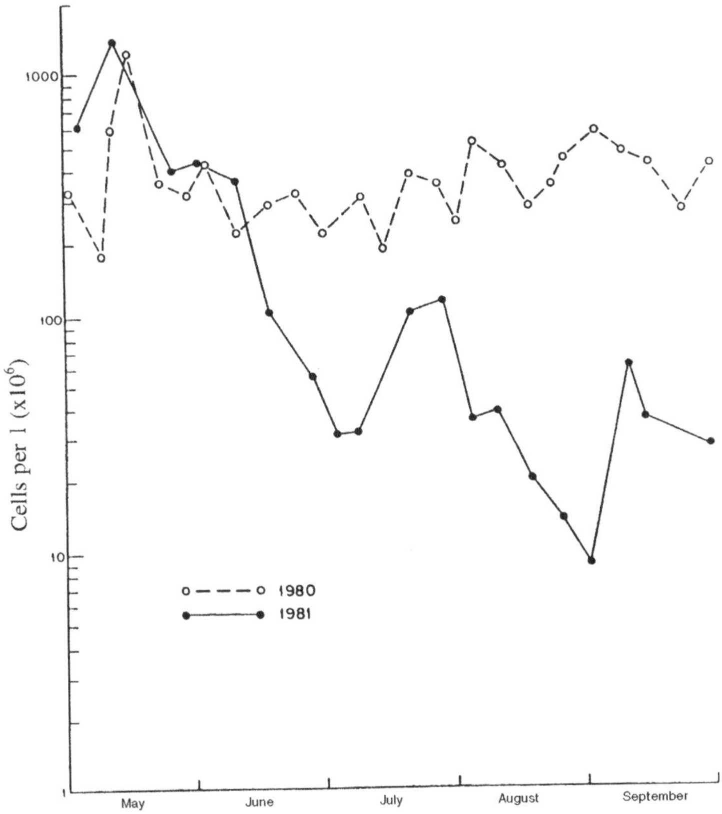

Fig. 7.2 Comparison in phytoplankton concentration, 1980 and 1981.

the fall. During the following year, the numbers of oysters being cultured in the pond were increased, and it was noted that phytoplankton abundance was considerably less than in 1980 (Fig. 7.2). It was considered possible that the carrying capacity of the pond was being exceeded and that surface water enrichment, utilizing the nutrient-rich bottom layer, might provide the means for improving oyster growth rate significantly. With improved growth rate, it might be advantageous to hold the oysters in the pond for two growing seasons rather than just one.

The phenomenon of upwelling, whereby surface water poor in nutrients is replaced by nutrient-rich deep water as a result of prevailing winds, is well-known in certain parts of the oceans[7.2]. The increase in nutrient availability in the presence of sunlight provides the stimulus for photosynthesis, resulting in high phytoplankton concentrations and, through the food chain, the basis for major fisheries. It seemed logical that bringing the nutrients from the bottom of the pond to the surface could have a similar impact.

With the assistance of a grant from the National Science Foundation, an experimental upwelling system was installed and operated during 1983. This consisted of an air compressor on shore connected, by means of a 390 m length of 1.9 cm plastic tubing, to an air diffuser located on the bottom at a depth of 6.5 m, in the deepest part of the pond. On the basis of calculations involving air compressor volume, tube length and diameter, and the depth and density of the water column above the diffusers, it was estimated that this air-lift system would lift water at a rate of approximately 6 m^3 per minute (Joseph Seales, New Alchemy Institute, personal communication). Major unknowns included the rate and extent to which the cold bottom water would become diluted with the surface layers, as well as the rate at which the high concentration of nutrients in the bottom layer would become depleted.

Five sampling stations were established in the pond to monitor temperature, salinity, pH, dissolved oxygen, phosphate, silicate, nitrate, nitrite, ammonia, phytoplankton and primary productivity. (Analyses for nutrient concentration, phytoplankton identification, and primary productivity were undertaken by Robert Silvia, Marine Research, Inc., Falmouth, Mass.). The location of these stations, as well as that of the diffuser, is shown in Fig. 7.3. The western area of the pond (West

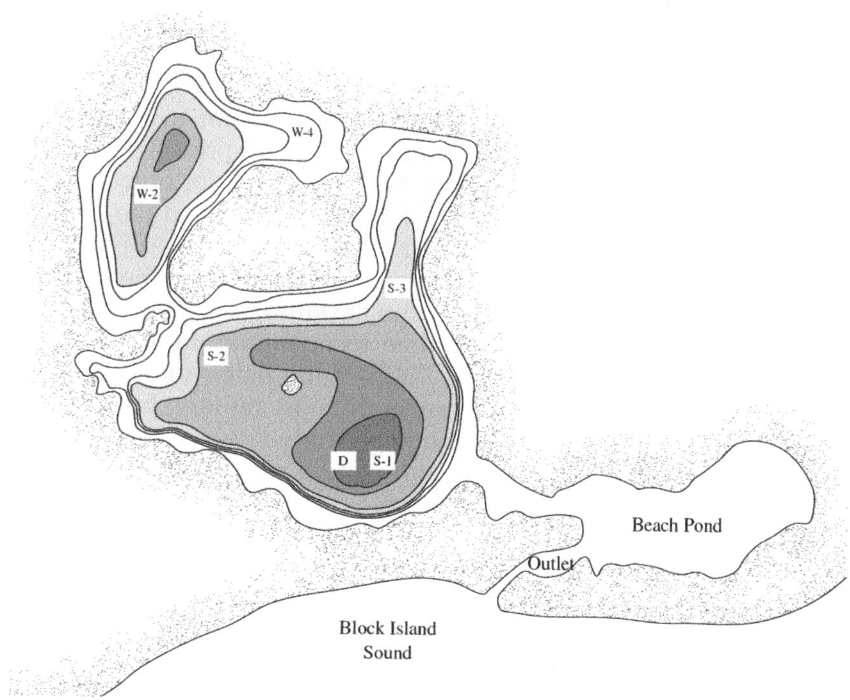

Fig. 7.3 Locations of the five sampling stations and diffuser (D).

Pond), nearly cut off from the southern area (South Pond) by a peninsula, was intended as the control area. Samples of 50 oysters were placed in lantern nets and suspended from buoys at each station for growth comparisons.

The upwelling system began operation in early June. By this date, the pond was strongly stratified. Hydrographic and chemical data in Table 7.2 comparing the results of observations at Station S-1 on 10 May and 31 May illustrate the seasonal increase in bottom nutrient concentrations as stratification intensified. When the air compressor was first turned on, the surface plume from the diffuser had appreciably higher concentrations of nutrients, specifically PO_4, SiO_3 and NH_3, than did the surface water at the regular sampling stations (Table 7.3). However, entrainment of water by the bubbles rising from the diffuser to the surface diluted the original bottom nutrient concentration by a factor of nearly three, as shown by comparing the concentrations in the surface plume with those in the bottom water at Station S-1 nearby.

It was also observed that nutrient concentrations in the bottom water fell rather rapidly during continuous operation of the air compressor. After 3 consecutive days of operation, stratification of the water column at Station S-1 had begun to break down. This was first indicated by the absence in the plume of the hydrogen sulfide odor characteristic of the bottom water. After 6 days of continuous operation, nutrient concentrations in the bottom water at Station S-1 had been reduced to roughly one-tenth the values recorded at the start of operations, while concentrations in the surface plume were only slightly higher than those at the other sampling stations.

Because of the efficiency of the upwelling system in mixing the pond water, it was decided to operate the air compressor in pulses, allowing enough time between periods of operation for nutrient concentrations to return to levels appreciably above those at the surface. Under this regime, operation was limited to an average of 14 days during June, July, August and September. Nutrient concentrations oscillated widely throughout the summer, falling within a few days after upwelling was initiated and then regenerating during the time the compressor was inactive (Fig. 7.4). Of the major nutrients measured, ammonium appeared to regenerate most rapidly, followed by phosphate and then silicate. Nutrient regeneration was usually accompanied by a fall in bottom temperature and a rise in bottom salinity as the water column restratified.

Despite the fact that the upwelling system succeeded in increasing surface nutrient concentrations appreciably in the immediate vicinity of the diffuser, its influence was not apparent at the sampling stations in South Pond. In fact, surface nutrient concentrations at Stations S-1, S-2 and S-3 averaged somewhat less throughout the summer than those at the control stations in West Pond (Table 7.4). This may have been due to several factors: rapid depletion of the nutrients near bottom, efficient vertical and lateral mixing of the upwelling bottom

Table 7.2 Temperature, salinity, pH, dissolved oxygen, and nutrient concentrations (µm) with depth; station S-1, May 1983

Date	Depth (m)	Temperature (°C)	Salinity (‰)	pH	O_2 (ml/ℓ)	Nutrient concentrations (µm)				
						PO_4	SiO_3	Nitrate NO_3	Nitrite NO_2	NH_3
10 May	0	13.5	16.2	8.0	9.5	0.15	17.61	3.35	0.32	2.83
	1	13.7	16.2	7.9	9.3					
	2	15.0	21.7	8.1	12.1					
	3	12.5	22.4	7.9	10.0					
	4	10.7	22.5	7.9	9.4					
	5	10.0	23.1	7.4	6.3					
	6	8.9	23.8	7.0	0.7					
	6.5	8.2	23.8	7.0	0.3	4.20	32.00	2.12	0.38	19.70
31 May	0	16.0	16.8	8.2	10.3	0.08	2.33	0.27	0.07	0.19
	1	16.0	17.9	8.2	11.0					
	2	15.7	18.2	8.4	12.5					
	3	15.8	18.9	8.6	16.7					
	4	15.5	20.3	7.6	7.6					
	5	14.3	21.2	7.2	4.0					
	6	10.5	22.4	7.0	0.3					
	6.5	9.5	22.6	6.9	0.3	9.56	49.43	0.00	1.65	140.00

Table 7.3 Surface nutrient concentrations (μm) at the diffuser plume and at other sampling stations, 7 June 1983

Station	PO_4	SiO_3	NO_3	NO_2	NH_3
S-1	0.06	5.40	0.66	0.11	3.48
S-2	0.06	3.19	0.33	0.10	1.70
S-3	0.05	7.86	0.60	0.13	6.49
W-2	0.08	8.22	1.65	0.10	1.75
W-4	0.07	11.89	0.79	0.10	3.20
Diffuser	2.36	22.00	0.00	0.07	59.22
S-1 (bottom)	7.32	58.99	0.00	2.52	132.00

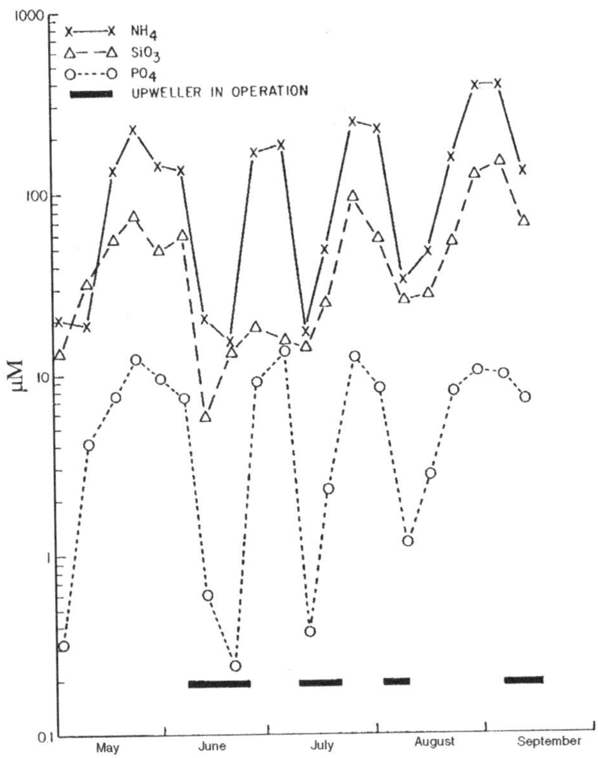

Fig. 7.4 Bottom nutrient concentrations at Station S-1 before and after operation of the diffuser.

water, and rapid assimilation of the nutrients by the phytoplankton as well as by macroalgae in the pond.

Monthly profiles of temperature, salinity, dissolved oxygen and pH at Station S-1 are depicted in Figs. 1.5–1.8. It is apparent that

Table 7.4 Mean monthly surface nutrient concentrations (μm) in the south (upwelling) and west (control) areas of the pond, 1983

Month	Number of samples	Pond	PO_4	SiO_3	NO_3	NO_2	NH_3
May	15	South	0.08	11.9	2.1	0.2	2.1
	10	West	0.08	12.2	2.1	0.2	2.5
June	12	South	0.09	6.1	1.1	0.1	1.2
	8	West	0.12	9.8	1.4	0.1	1.9
July	12	South	0.06	7.7	0.3	0.1	0.4
	8	West	0.05	11.0	0.9	0.1	1.4
August	15	South	0.10	7.4	0.7	0.1	1.1
	10	West	0.11	13.2	1.2	0.1	0.8
September	12	South	0.07	9.9	0.4	0.1	0.4
	8	West	0.20	11.9	0.3	0.1	0.7

Table 7.5 Mean rate of carbon uptake (μg C/ℓ/day) each month, 1983

Month	S-1	S-2	S-3	W-2	W-4
May	42	58	69	69	73
June	328	226	259	127	254
July	132	97	207	261	174
August	138	136	117	119	140
September	328	424	483	442	554

hypoxic and acidic conditions in the subsurface layers steadily rise from the bottom towards the surface as the season progresses, reversing in late September after breakdown of the thermocline and halocline.

The results of carbon-14 analyses were not conclusive, with a high degree of variability between replicate samples. As indicated in Table 7.5, rates of carbon uptake in South and West Ponds were generally similar.

Observations on phytoplankton population densities during the period May–September are summarized in graphical form in Fig. 7.5. In order to simplify the results, the graph representing the population in South Pond is based upon the weekly average of the samples collected at Stations S-1, S-2, and S-3 twice each week. Similarly, the graph for the populations in West Pond, the control area, represents the average of the samples collected at Stations W-2 and W-4. The periods of air compressor operation are also indicated.

Two features of these graphs seem noteworthy. First, the population densities of phytoplankton were frequently found to be considerably

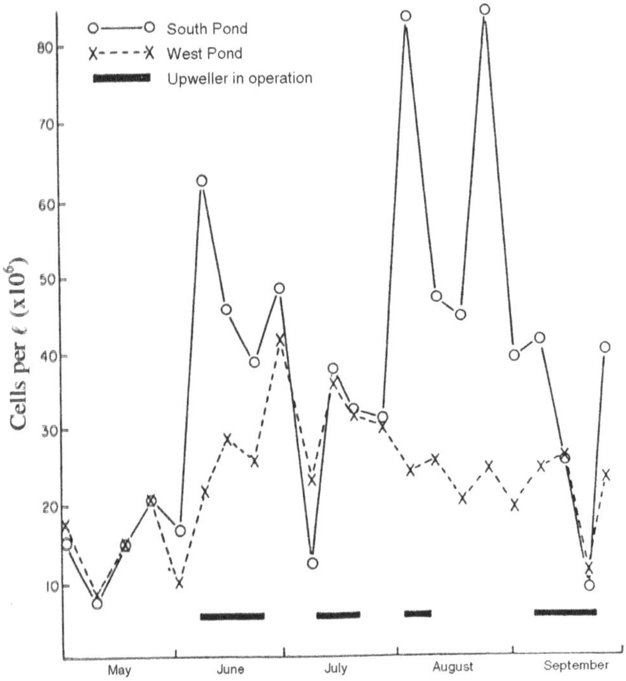

Fig. 7.5 Comparison of mean phytoplankton concentrations at Stations W-1 and W-2 and at Stations S-1, S-2, and S-3.

higher at the three stations nearest the diffuser than at those stations in the control area (West Pond). By integrating the areas under both curves, it is estimated that the standing crop of phytoplankton in the control area over the 5 month period was only 61% of that for the fertilized area (South Pond). This is in contrast to results obtained during the preceding year, when phytoplankton samples were collected at the same five stations at similar frequencies and when the upwelling system was not in operation. During that year, it was estimated by similar integration that the standing crop in West Pond was about 88% of that in South Pond.

Also of interest is the relationship between the sharp peaks in abundance of phytoplankton that occurred in South Pond during the summer and the period during which the air-lift system was operating. The first major population peak occurred in early June, most markedly at Station 1, nearest the diffuser, 3 days after the compressor was turned on. The dominant species were small flagellates and diatoms, primarily *Cyclotella* spp. and *C. meneghiniana*. A second significant bloom, consisting primarily of the diatom *Skeletonema costatum* and small flagellates, occurred during July, also 3 days after the

Table 7.6 Diatom densities (cells per $\ell \times 10^6$) at stations S-1 and W-4 immediately before and after operation of air compressor (air compressor turned on 7 June, 9 July, 2 August and 6 September)

Date	S-1	W-4	Dominant Species
7 June	3.68	3.54	
10 June	20.00	2.48	*Cyclotella* spp.
6 July	5.10	7.55	
12 July	16.70	8.06	*Skeletonema costatum*
2 August	11.60	2.43	
5 August	25.90	1.00	*Cylindrotheca closterium* *Thalassiosira pseudonana*
6 September	6.96	8.86	
9 September	27.80	11.00	*Cylindrotheca closterium*

compressor was reactivated. A third major peak was observed in August, again 3 days after compressor reactivation and dominated largely by the diatom *Cylindrotheca costatum* as well as small flagellates. Finally, 3 days after the compressor was turned on in September, a second bloom of *Cylindrotheca* occurred. As shown in Table 7.6, there was no such correspondence between diatom blooms and activation of the air compressor at the station most remote from the diffuser (i.e. Station W-4).

A total of 79 different species of phyoplankton were identified at the five stations during the 5 month period. The populations in the two areas of the pond were generally similar with respect to species composition. However, in terms of both absolute and relative abundance, diatoms in general were significantly more dominant in the area of upwelling (South Pond). While the diatom populations tended to fluctuate sharply during the 5 month period in terms of both abundance and species, unidentified microflagellates were consistently abundant at all stations throughout the period and were always one of the dominant forms, if not the dominant form. Similar results with respect to the species composition of phytoplankton were reported for a brackish water estuary in Long Island, New York[7.3]. The relative abundance of the six phytoplankton groups, based upon total cell counts at each station during the 5 month period, is indicated in Fig. 7.6.

The average rate of growth (shell height) of the oysters held in lantern nets at Stations S-1, S-2 and S-3 are compared with that for the oysters at Stations W-2 and W-4 in Fig. 7.7. Statistically, the differences in mean shell height at the five stations at the termination of the growing period were found to be highly significant, with growth clearly superior in South Pond. Additional data relative to growth are given in Table 7.7. Total volume of the sample oysters held at the South Pond stations increased on the average by a factor of nearly 14

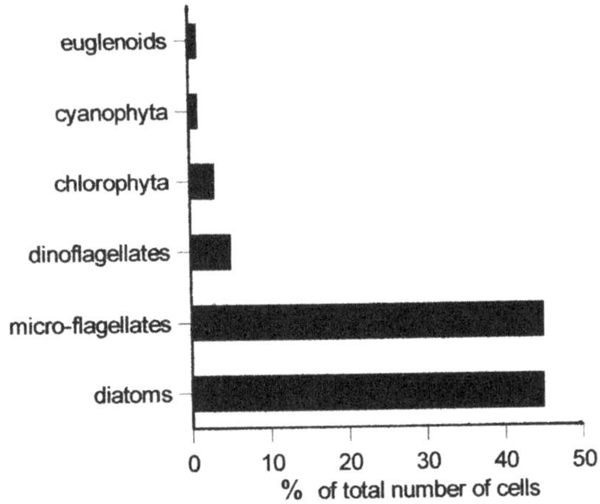

Fig. 7.6 Relative abundance of different classes of phytoplankton in Island Pond.

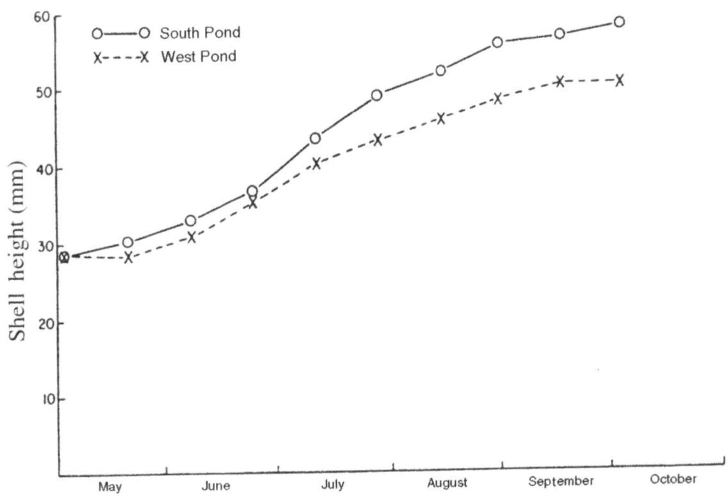

Fig. 7.7 Comparative oyster growth rates in South Pond and West Pond.

during the 5 month period, compared with a factor of 9 for the oysters in West Pond. Similarly, dry meat weights increased by a factor of nearly 37 in South Pond, compared with a factor of 26 in West Pond. Since South Pond is more exposed to winds than West Pond, it is possible that wind-induced currents would tend to favor growth rate in the former area. However, similar growth rate experiments undertaken the preceding year failed to indicate a superior rate of growth in South Pond.

Table 7.7 Oyster growth parameters at different sampling stations, 1983

Date	Station	Sample size (N)	Mean shell height (mm)	Mean total volume (ml)	Mean shell volume (ml)	Mean dry meat weight (g)	Condition index
3 May 1983	control group	50	27.5	1.4		0.03	
30 Sep 1983	S-1	52	57.8	19.3	8.5	1.05	9.7
	S-2	52	58.9	19.2	7.9	1.07	9.5
	S-3	46	58.5	19.4	8.0	1.18	10.4
	W-2	51	50.3	12.5	4.8	0.75	9.7
	W-4	49	50.0	12.6	5.2	0.80	10.8

To summarize the results of this experiment, the abundance of phytoplankton, believed to be the primary source of nutrition for oysters, was significantly higher in the area of the pond with the upwelling system than in the control area. This was not the case during 1982, when phytoplankton data indicated that the standing crop of phytoplankton in West Pond over a similar period was about 88% of that in South Pond (compared with only 61% in 1983). Also, oyster growth rate was significantly greater in the upwelling area of the pond than in the control area, which had not been the case for growth rate experiments during 1982, when growth rate in the former area was actually slightly less than in the latter.

However, it was evident that the observed increases in the standing crop of phytoplankton in the test area occurred in pulses and could not be sustained by continuous upwelling. Furthermore, even though there was evidence that oyster growth rate did benefit from the upwelling system, the difference in growth between the groups of oysters in the area of upwelling and those in the control area did not seem sufficiently great to justify a modification of the culture system, i.e. holding yearling oysters in the pond for a second growing season.

8 Oyster Culture in Western Europe

The natural oyster resources of Western Europe have historically played a significant role in the fishing industries of this region. Unfortunately, the rich natural beds of oysters in Great Britain and on the continent were highly vulnerable to excessive and unregulated exploitation. With increasing populations and industrialization, levels of pollution increased as well. Towards the end of the nineteenth century, when oyster production in the United States was near its peak, the industry in parts of western Europe was already in decline.

The best known and most widely distributed species of oyster in Western Europe has historically been the European flat oyster (*O. edulis*). The culture of this species dates as far back as the first century BC, when the Romans utilized small, brackish-water lakes near Naples, Italy, for this purpose. The range of the flat oyster extends from the Mediterranean Sea to Norway, where isolated populations occur north of the Arctic Circle[8.1]. This is not an estuarine species, preferring the relatively cool and saline waters of open bays and exposed shore lines.

Until relatively recently, a species of cupped oyster, *Crassostrea angulata*, also played a significant role in the oyster industry of Western Europe. Usually referred to as the Portuguese oyster, this species was much more restricted in its distribution than the flat oyster, occurring primarily along the coasts of Southern France, Portugal, Spain and North Africa. Unlike the flat oyster, it does not reproduce at temperatures below 20°C, and hence its relatively narrow distribution along the European coast. The French industry was changed radically in 1868 when an entire cargo of Portuguese oysters was dumped overboard near the mouth of the Gironde River during a storm[8.2]. Well suited to these particular estuarine conditions, the oysters became successfully established and ultimately succeeded in colonizing the French coast line as far north as the mouth of the Loire River. In so doing, this species became a valuable supplement to the dwindling stocks of flat oysters.

During the past few decades, the abundance of both these species has drastically declined following the occurrence of epizootic diseases. In an effort to salvage local oyster industries, the Pacific oyster, *C. gigas*, was imported to several European countries from Japan and the Pacific Coast of North America. This introduction was successful to the extent that certain of the oyster industries of Western Europe still survive and now rely heavily upon this species. The flat oyster, while still in high market demand, has become increasingly scarce, while the Portuguese oyster has all but disappeared.

The cause of the epizootic disease that first afflicted the Portuguese oyster populations during 1967–1970 was initially unclear but was linked to abnormal lesions of the gills. The condition became known as 'gill disease'[8.3]. This was followed soon after by an even more devastating disease that virtually destroyed the Portuguese oyster industry in France[8.4]. Subsequent studies have implicated two viruses, the gill necrosis virus (GNV) and the hemocytic infection virus (HIV), as the probable causes of these epidemics[8.5]. In the case of heavy mortalities among European flat oysters that began to occur in France in 1968 and again in the late 1970s, the parasites involved in both cases were both identified as protozoans (*Marteilia refringens* and *Bonamia ostreae*). The latter pathogen subsequently became widespread throughout much of Europe, e.g. Spain, the Netherlands, Denmark and the United Kingdom. The oyster industries in these countries, traditionally reliant upon the flat oyster, have never really recovered. Only France continues to produce large quantities of oysters, about 92% of total European production in 1995, thanks to the adaptability and hardiness of the Pacific oyster. This species appears to be resistant not only to the two viruses that destroyed the Portuguese oyster populations but also the two protozoan parasites, *Bonamia* and *Marteilia*[8.6].

In recent years, France has ranked number five in world oyster production. In 1995, a total of 148 000 t of oysters were produced in France, nearly 90% of which can be accounted for by the Pacific oyster (Fig. 8.1). Importations of this species, initiated on a small scale in 1968 and continued on a much larger scale in 1969 and 1970, resulted in the establishment of a dense and self-sustaining population along France's Atlantic Coast extending from Arcachon northward to the mouth of the Loire River[8.7]. Less than 2% of landings today consist of flat oysters, while the Portuguese oyster accounts for about 8%. In the opinion of some oyster taxonomists, the Pacific and Portuguese oysters are so similar as probably to be the same species with some slightly different characteristics, including degree of susceptibility to disease[8.8].

The basic system of oyster culture practiced in France today was initiated around the middle of the nineteenth century, when it became apparent that the natural oyster beds had been seriously overexploited, and in certain respects has changed very little[8.2]. The major producing areas lie along the Atlantic coast between Normandy and the Bay of Arcachon, roughly 150 km north of the Spanish border. Some oyster culture takes place in the Mediterranean Sea, but production is small, and the growers rely completely upon the oyster producers along the Atlantic coast as their source of seed[8.9].

The majority of the oysters produced in France are grown intertidally. This is made possible by the very large expanse of tidal flats in many of the protected coastal areas suitable for culture, as well as

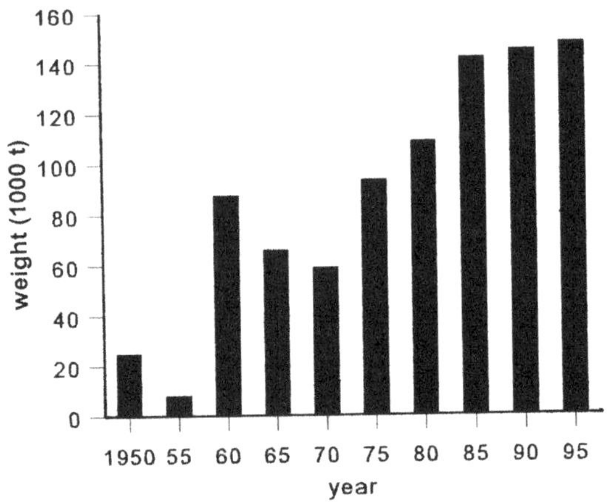

Fig. 8.1 Oyster production (whole weight) in France.

by the absence of ice and prolonged periods of freezing weather that would destroy oysters on exposed beds. In the bays of Oleron and Arcachon, where intertidal culture is practiced, the tidal amplitude may exceed 3 m under spring tide conditions. In the northern part of France – Normandy and the north coast of Brittany – many of the oysters are cultured subtidally, and total production from these areas is relatively small. Because of low water temperatures along this part of the French coast, natural reproduction of the Pacific oyster rarely occurs, and the seed is imported from the beds in Arcachon, Marennes-Oleron and the Gironde estuary or purchased from hatcheries[8.9].

The history of the French oyster industry is an interesting one. By the mid-nineteenth century, a large percentage of the natural beds along the French coastline had been virtually depopulated by overfishing. The French embryologist V. Coste, impressed by simple culture techniques used near Naples, undertook to apply these methods in France, beginning in 1857[8.2]. At selected sites, intertidal bottom was hardened by spreading stones or shell, mature oysters were collected and planted on the surface of this substrate, and sticks were assembled in bundles and placed on the intertidal flats as cultch. These initial experiments yielded highly encouraging results. It was soon found that semi-cylindrical roofing tiles provided a more efficient and durable type of cultch material than faggots. However, many oysters were destroyed in the process of dislodgement. Further experiments revealed that, if the tiles were dipped initially in a mixture of lime and sand, the oysters readily attached to this surface and, as juveniles, could be easily flaked off without injury to their shell. This type of collector is still commonly used in France, nearly a century after the technique was first developed.

Fig. 8.2 Oyster parcs, Arcachon, France (photo by O. H. Matthiessen).

In Arcachon, the tiles are approximately 0.5 m in length and 15 cm wide. These are dipped in a lime-sand mix and allowed to dry before use. Other forms of cultch include plastic (PVC) pipe, about 2 m in length, the outer surface of which has been grooved and roughened in order to make it more attractive for the oyster larvae, plastic mesh bags filled with oyster shell, slabs of slate, and even scrap iron. In some areas, empty shells are strung on galvanized wire and suspended off the bottom from fixed supports. Many growers prefer the tiles, however, since the lime coating makes it possible for the oysters to be flaked off without damage or injury.

The Pacific oyster normally spawns during July when water temperatures reach about 23°C[8.10]. Although, in the Bay of Arcachon, spatfall is quite predictable, the abundance and stage in development of the oyster larvae are monitored carefully by government biologists, who advise the oystermen as to the optimum time for setting out the cultch on their beds. The great majority of Pacific oysters produced in France are cultured, initially at least, upon carefully defined areas of intertidal ground leased from the government, referred to as 'parcs' (Fig. 8.2). In the Bay of Arcachon, the average size of a parc is about 700 m^2 in area, and one family may control several. The borders of each parc are marked by stakes (Fig. 8.3), driven into the sand to form fences. These help exclude finfish predators as well as reduce the velocity of current flow over the beds, which otherwise would scatter the oysters.

Fig. 8.6 Oyster claires, Oleron, France (photo by O. H. Matthiessen).

When the oysters are nearly 2 years of age, they are removed from the plastic bags and sorted by size. The largest oysters, 75 mm or more in size, are marketable. The smaller oysters are either spread on the bottom or placed back in bags for further growth.

In the Marennes-Oleron area, where the greatest number of oysters in France are produced, the system of culture includes the extensive use of 'claires'. The claires are shallow salt water ponds that dot the lowlands of this area (Fig. 8.6). These ponds communicate with the sea by means of channels controlled by tidal gates and were originally constructed as evaporation lagoons for salt production. The salt industry had declined by the early nineteenth century, and the abandoned claires were soon utilized for the growing oyster industry. They are presently privately owned, although many are leased to oyster growers.

The average claire is about 40 m long and 12 m wide, with a minimum depth of 0.5 m. The bottom consists of clay. In the spring, the claires are usually drained dry during a low spring tide, exposing the bottom to the air and allowing it to dry. For reasons not clearly understood, this treatment of the sediment makes available certain nutrients, presumably bound to the sediment, that favor the growth of a particular species of microscopic plant (*Navicula ostrearia*) when the claire is refilled with sea water after a few weeks. This planktonic algae imparts a greenish color as well as desirable flavor to the oyster's flesh.

Marketable oysters of superior size and shape are brought from the parcs and placed on the bottom of the claires at low densities, about ten per m^2, where they remain throughout the summer. It is by no means certain that an algal bloom, most particularly of the diatom *Navicula*, will occur. If it does, and the greenish coloration develops, the oysters are immediately harvested, which is usually during September, and sold at a premium price. Additional oysters are then brought from the parcs to the claires, and, although planted at much higher densities, these oysters may fatten and also acquire a greenish tinge, thus improving their market value.

The use of claires is principally in the basin of Marennes-Oleron. The sector of coastline immediately to the north and extending from the mouth of the Loire River to the mouth of the Charente, and referred to as Re-Centre Ouest, is second only to the basin of Marennes-Oleron in terms of volume of oyster production. Here oyster spat are collected on grooved tiles, on oyster shells threaded on rods and laid horizontally on metal racks in the intertidal zone, and on plastic pipe. Although a percentage of these oysters are finished off for market, many are sold at an age of 18 months to private growers in Brittany and in other areas of France where the waters are either too cold or the coastline generally unsuitable for the successful reproduction and settlement of this species.

The oyster industry along the Mediterranean coast of France is quite small, about 5% of national production. These oysters, mostly *C. gigas*, are usually obtained from Arcachon, Marennes-Oleron or Re-Centre Ouest, or even from Japan, and may arrive still attached to the shell used as cultch. Rods are inserted through the holes pierced through the shells and are suspended from racks. After 9–10 months of growth, the oysters are detached from the shells and fastened, individually, by means of quick-setting cement, to lengths of wood, which are laid horizontally on intertidal racks.

The Pacific oyster has come to play a dominant role in the French oyster industry within a relatively short period of time. Furthermore, it has reproduced so successfully along the southern part of France's Atlantic Coast that there is some concern that overcrowding in several of the major producing areas, such as Arcachon and the Bay of Oleron, may be imposing a severe stress upon these populations. Apparently, oysters in the Bay that used to require only a period of 18 months in which to reach a desirable marketable size now may require up to 3–4 years. For this reason, the stocking density in the Bay of Marennes-Oleron, which has the highest level of production in Europe (40 000–60 000 t annually) has been limited to 6000 oyster bags, or 72 t, of oysters per hectare[8,11].

The majority of flat oysters produced in France today are grown along the south coast of Brittany. The Gulf of Morbihan is the primary area for the collection of spat, with limed ceramic tile the principal cultch material used. The tiles, with holes drilled through each end, are wired together in bundles of 10 or 12, called 'bouquets'. The bundles are then dipped in a hydrated lime-sand mix and dried.

In this region, the flat oyster generally spawns when water temperatures in summer rise above 15°C, and setting may occur from June to September. The occurrence of oyster larvae in the water is carefully monitored by government biologists. When a sufficient number of larvae is detected, the bouquets of limed tiles are stacked in the intertidal zone on metal supports or treated wood to prevent them from sinking into the mud.

The young oysters are not removed from the tiles until the following spring. At this time they are spread on the bottoms of intertidal parcs, which have often been hardened by the addition of sand. Planting density in some areas has been reduced from 0.5 to 0.1 kg/m^2 in order to increase growth rate and attain marketable size before losses to the parasite *Bonamia*[8.9]. Plastic mesh and stake fencing around the borders of the parcs help exclude predators such as crabs and finfish and prevent tidal currents and surge from washing the small oysters away. Here the oysters remain for one year before being collected, either by dredge during high tides or by rake or hand during low tides, and transferred to other growing areas. Many of these 18 month oysters are trucked to Northern Brittany or other parts of France for further growth. Before the occurrence of *Bonamia*, many were sold to private growers in Holland or Spain. Oysters transported to northern Brittany are generally cultured for market in relatively deep water and are harvested by bottom dredge.

Shellfish hatcheries do not play a large part in France's oyster industry, largely because natural reproduction of Pacific oysters is sufficiently dependable. Also, the price of hatchery seed is considerably higher than seed derived from natural sets. Attempts to produce significant quantities of the flat oyster by hatchery methods have been abandoned because of the likelihood of infection of the seed by *Bonamia*.

Compared with France, oyster production in the other European countries is very small. In Great Britain, the oyster fishery was well on the wane before the turn of the nineteenth century. The heart of the industry lay near the mouth of the River Thames, and, in 1864, something in the order of 500 000 000 oysters were sold in the London market of Billingsgate[8.2]. The decline began with excessive exploitation of the natural beds and increasing pollution. It was hastened further by the accidental importation of two serious oyster pests: the slipper shell (*Crepidula fornicata*), which rapidly spread over the beds, creating soft muddy conditions intolerable for oysters, and the predatory oyster drill (*U. cinerea*). During World War I, the oyster beds lay unattended, allowing further ruination by *Crepidula*[8.12], and this was followed by an epizootic disease, which destroyed large numbers of oysters during the period 1919–1923 in Great Britain and parts of Europe. Production fell from over 33 000 000 oysters in 1912 to less than 17 000 000 by 1924. (Between 1950 and 1995, annual production has averaged only 380 t, or roughly 5 000 000 oysters per year).

Research on the breeding of oysters under controlled conditions, in this case large outdoor concrete tanks, was initiated at the Fisheries Research Station in Conwy during the 1920s in an effort to salvage the industry. Although the results of these early experiments were erratic in terms of production, useful facts relating to the biology of oysters were discovered, including the determination that oyster larvae could not efficiently utilize phytoplanktonic food greater than 10 μm (about the size of a red blood cell), and that small flagellates with thin cell walls composed the most satisfactory diet. These observations provided basic information for the subsequent development of commercial hatchery techniques[8.13]. Research at Conwy on various aspects of oyster culture has continued, but the industry has never returned to its former status, owing in large part to pollution and consequent closure of oyster beds.

The oyster industry of Great Britain historically was based upon the harvest of natural stocks of the European flat oyster. As this species declined in abundance, Portuguese oysters were brought over from the continent each year as a supplement to the dwindling native stocks. The water temperatures of Great Britain, however, are too cold to allow this species to reproduce, and its own decline as a result of disease during the late 1960s and early 1970s terminated further shipments. It was decided by 1965 to introduce the Pacific oyster, initially on an experimental scale, since this species was reported to thrive at water temperatures very similar to those characteristic of Great Britain[8.14].

What little oyster production remains in Great Britain is presently based upon limited natural stocks of flat oysters, mostly from the Solent, and planted beds of hatchery-reared Pacific oysters, which are able to grow at Britain's prevailing water temperatures but do not reproduce successfully other than in a few sheltered areas where water temperatures may reach 23–24°C. Juveniles raised in the hatchery are usually held in upwelling systems until 10 mm or so in size and are then transferred to plastic mesh bags at densities of about 2500 oysters (2.5 kg) per bag. The bags are supported horizontally off the bottom on racks in the intertidal zone. In areas of good water flow and high plankton concentrations, the bags may contain up to 20 kg of 20 g+ oysters. Depending upon the site, a market size of 70–75 g is reached in 3–4 years[8.15]. In England, oysters must be depurated prior to marketing, which is generally accomplished by holding the oysters in tanks, filled with water sterilized by ultraviolet light, for 48 hours.

The oyster industry in Ireland has produced about 1600 t per year on the average during the 10 year period 1986–1995. This industry also has been traditionally based upon the harvest of the flat oyster from natural populations, supplemented to some extent by eastern oysters imported from the United States. The latter species was unable to establish natural populations in Irish waters, since water temperatures do not generally become high enough for successful reproduction. Most of the flat oyster population in Ireland, as in much of Europe, were destroyed by disease during the 1920s[8.16], and this fishery has had difficulty recovering. Natural beds have been depleted, and natural reproduction is no longer sufficient to assist the industry appreciably. Nevertheless, there has been a modest increase in flat oyster production in recent years, from 388 t in 1986 to over 1200 t in 1995.

Populations of *O. edulis* on the west coast of Ireland generally spawn when water temperatures reach approximately 15°C in early summer[8.17], and the cultch material is put out during June and July. Various types of cultch material are used, including shell and lime-coated plastic[8.18]. The oysters remain on the cultch until early in the following spring, when, at a size of about 10–15 mm, they are removed and placed in plastic mesh bags, or in trays suspended from the surface, or are simply scattered on the bottom in intertidal parcs. It generally requires 3–5 years for the oysters to reach a marketable size of 65–70 g.

The Pacific oyster was first introduced to Ireland in 1969[8.19]. Because of Ireland's relatively low water temperatures, very little natural recruitment occurs, and this phase of the industry relies upon hatchery production of seed. However, rapid growth has enabled this species to reach marketable size in 2 years, and it appears to be resistant to the diseases that have afflicted western Europe. Most Pacific oysters are grown in plastic mesh bags supported off the bottom on racks or spread on the bottom. During the past 10 years, production has increased from 120 t in 1986 to over 2500 t in 1995, or about double the volume of flat oysters harvested that same year.

Long a producer of *O. edulis*, the Netherlands also has witnessed a gradual replacement of one species by another in its oyster industry. The stocks of flat oysters, upon which the industry was based entirely, were nearly eliminated by abnormally cold weather conditions during the early 1960s. This, combined with government plans to permanently seal off the Oosterschelde – a large estuary draining into the North Sea and the Nation's primary shellfish culture area – from the sea, drove the majority of oystermen out of business[8.20]. Even though this 'Delta Plan' was ultimately modified to allow some circulation into the Oosterschelde, it was soon discovered that very few strains of this species, imported from elsewhere to supplement the remaining indigenous stocks, could survive the cold winters in this area.

Nevertheless, the industry survived by importing large seed oysters from France and England. These oysters, obtained in the spring, remained on the beds for only one growing season before being harvested, in order to avoid winter mortalities. However, this industry came to a halt with the occurrence of the pathogen *Bonamia* in 1980. This disease was introduced with the imported seed oysters from France, resulting in high mortalities. As a result, no further importations were permitted.

At the present time, a limited number of seed oysters are obtained from Lake Grevelingen, a salt water lake that contains a small natural population of flat oysters[8.21]. These oysters have reproduced regularly and appear to be tolerant of *Bonamia*. Furthermore, the seed are sufficiently hardy to withstand the majority of winters characteristic of this region. However, total production has been reduced to only a few million flat oysters each year.

Although the oyster farmers in Holland concentrated upon the production of flat oysters in the past, the focus has now shifted to the Pacific oyster, which was introduced on a trial basis in 1964 as a substitute for the fast-disappearing Portuguese oyster. The hardy Pacific oyster has managed to reproduce naturally in spite of relatively low water temperatures. The spat are collected on mussel shells, which are spread on bottom plots in shallow water and gathered by hand when of marketable size[8.21] During the period 1990–1995, the annual harvest averaged 1500t.

The oyster industry of Spain has never been particularly significant in relation to its large sea mussel fishery. For the most part, it has been concentrated along Spain's north-west coast, most particularly in the rias, or inlets, of Galicia, where water temperature and quality at one time favored the natural reproduction and recruitment of flat oysters.

Depletion of natural beds led to dependence upon imported seed, primarily from France. Something to the order of 600t, or about 100000000 18 month old oysters, were imported annually[8.22]. However, the destruction of large numbers of oysters in France by disease, beginning in 1968, terminated the source of seed. Oyster production in Spain fell from 2300t in 1973 to only 8t in 1979. Since then, annual production has increased somewhat. In 1995, over 5000t, consisting mostly of flat oysters, were landed.

Traditionally, the spat were collected on tiles held in racks, positioned in the intertidal zone, in early summer[8.23]. Spawning first occurred in early May, when water temperature had risen to 15°C or more, and in some years would continue throughout the summer until September. The tiles, coated with a thin layer of lime, were brought ashore at the end of summer, when the spat were scraped off into shallow trays, constructed of wooden frames and galvanized wire, supported about 20cm above the bottom. Growth in the rias was extremely rapid owing to favorable temperatures and water rich in phytoplankton, and the oysters would attain a size of 60mm within 9 months.

The majority of Spanish oysters today are produced by suspension methods[8.24]. Seed oysters are imported from Italy, Greece, Egypt and

England. Some growers cement their seed (50 mm in size) to ropes, about 5 cm apart, and the ropes are suspended from rafts. The oysters are harvested after 9–10 months at a size of about 80 mm. Fouling of the ropes by mussels can be intensive, and infection by the parasites *Marteilia* and *Bonamia* remains a constant threat. Some Pacific oysters, raised in hatcheries in France and the United Kingdom, have been cultured to market size in Spain. However, production has been limited by poor market acceptance[8.24].

9 Working Around Disease

In the early spring of 1985, very heavy mortalities were discovered among seed oysters in Cotuit Bay on Cape Cod. Some of these oysters had been produced by Ocean Pond Corporation and planted on the Cotuit Oyster Company's beds the previous fall. Pathological examination of the surviving oysters indicated these to be heavily infected with the protozoan parasite MSX (*H. nelsoni*). This parasite had never been detected in samples collected each year from Island Pond population, and it was concluded from previous pathological examinations that in all likelihood it had been introduced with oysters obtained from the Hammonassett River in Connecticut. By the end of the summer of 1985, total losses on the Cotuit beds were estimated at 85%.

H. nelsoni first appeared in lower Delaware Bay in 1957. Within 3 years, it had destroyed about 90–95% of the oysters on the planting grounds and more than half the oysters on the seed beds[9.1]. By the spring of 1959, this parasite had invaded the Chesapeake Bay and proceeded to destroy 90–95% of the oysters in the lower Bay within a 3 year period. During the past 40 years, MSX has worked its way northward along the Atlantic Coast, eventually invading oyster beds in parts, if not most, of southern New England. In 1985, however, the occurrence of MSX along the south shore of Cape Cod had never been recorded, and therefore virtually nothing was known about the seasonality of infection and subsequent mortality in this region.

With support from the Massachusetts Sea Grant Program, investigations were initiated in 1986 to determine the seasonal pattern of MSX infection and subsequent mortality in Cotuit Bay[9.2]. In April, a group of MSX-free oysters was obtained from Island Pond and planted on the Cotuit beds, thereby exposing them to infection. Samples of oysters were collected from this initial group each month and analyzed for the presence of MSX. Additional samples of oysters from Island Pond were brought to Cotuit each month, May–October, and the mortality rate in each group was compared.

These initial observations indicated a relatively short period of infection during mid-summer, i.e July–August, suggesting that oysters introduced to an infected area after August might escape infection until the following summer. A possible solution to the problem of transplanting oysters from Island Pond to infected areas, therefore, would be to transplant after August, when the likelihood of infection appeared to be small, with the intention of harvesting before the end of the following summer, when a new cycle of infection would begin. Since the period between initial infection with MSX in Cotuit Bay and death was determined to be a minimum of 2 months, the relatively few oysters that might become infected after August might nevertheless survive to marketable size during the fall.

At this time, as discussed in an earlier chapter, Ocean Pond was under contract with the New York State Urban Development Corporation to evaluate the technical and economic feasibility of growing oysters on a commercial scale in Island Pond for more than 1 year. The original purpose of this study had nothing to do with disease, but the results made it quite clear that attempting to work around the disease problem in Cotuit Bay by holding the oysters in Island Pond for an additional growing season had small chance of success. The basic problem was that the oysters being cultured in the pond at this time, and being used for experimental purposes, were the progeny of wild pond oysters and simply did not reach a large enough size in the pond during their second season to circumvent the disease problem in Cotuit Bay. In effect, they would not become marketable on the Cotuit beds soon enough to avoid infection and subsequent mortality.

The possibility of holding oysters over for a third growing season in the pond was also explored. The great majority of these oysters held over during their third growing season were in fact large enough to be planted on the Cotuit beds with the likelihood of survival to market. In order to achieve this, however, and because of spatial requirements for the oysters, about 2000 five-tier lantern nets would be required for a single year-class of 1 000 000 oysters. This approach did not seem practical for Island Pond, given the number of nets that would be required and the evidence that attempting to grow significant numbers of oysters in the pond even for a second year might be straining the pond's capacity for food production.

As an alternative approach to the disease problem, it was decided to test a strain of oysters that were genetically resistant to MSX[9.3,9.4]. This project was funded in part by a grant from the National Science Foundation. In the spring of 1987, a small group of such oysters was obtained from David Relyea, manager of the Frank M. Flower Oyster Company in Bayville, NY. As discussed in Chapter 5, these oysters were members of the sixth generation of a strain originally collected from Long Island Sound in 1965 and bred for resistance at the Rutgers University Shellfish Research Laboratory on Delaware Bay. These oysters were induced to spawn at the Ocean Pond hatchery facility, and the progeny were reared in Island Pond during their first growing season.

In July 1988, groups of oysters of the resistant strain (LISR, or Long Island Sound Resistant) and of the wild pond population (FIS, or Fishers Island Susceptible) of a similar age and size were planted in trays on the Cotuit beds, with the intention of comparing their growth,

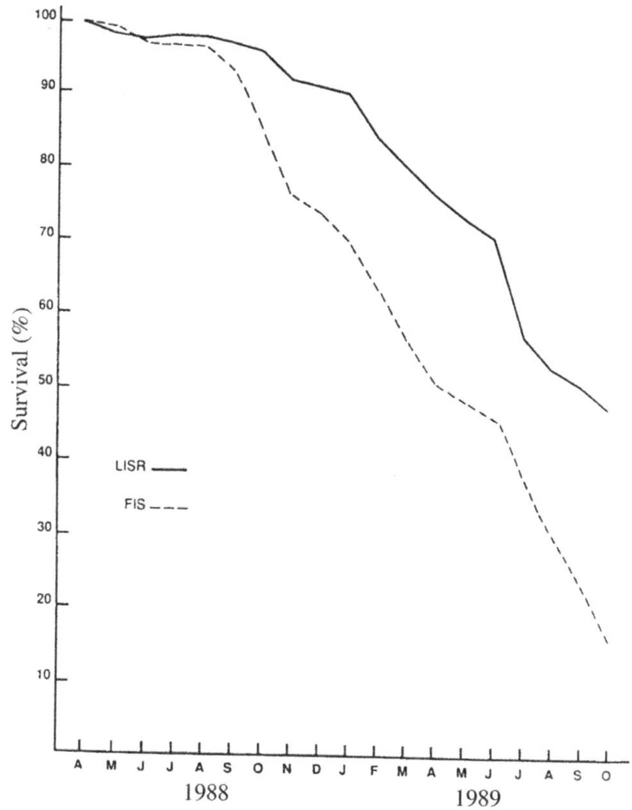

Fig. 9.1 Relative survival of resistant (LISR) and susceptible (FIS) oysters in Cotuit Bay, Massachusetts.

susceptibility to MSX infection, and mortality rate under MSX pressure. By October, infection among the FIS group exceeded 90%, or nearly double that for the LISR oysters (56%)[9.5]. Mortality at this time exceeded 20% for the FIS group but was less than 10% for the LISR group (Fig. 9.1). Furthermore, with respect to both shell height and total weight, the FIS oysters had fallen significantly behind the resistant group (Figs. 9.2 and 9.3). By April 1989, nearly 50% of the LIS group had died, as compared with less than 25% for the LISR oysters. In October of the same year, total mortality among the FIS oysters was nearly 85%; in the LISR group, total mortality approximated 50% (Fig. 9.1)[9.5].

By November 1988, nearly 30% of the resistant oysters had reached the minimum marketable size of 75 mm. None of the oysters of the susceptible strain, on the other hand, had done so. If the resistant oysters had been spawned early in the spring, as is often the case in commercial hatcheries, it is likely that a much higher

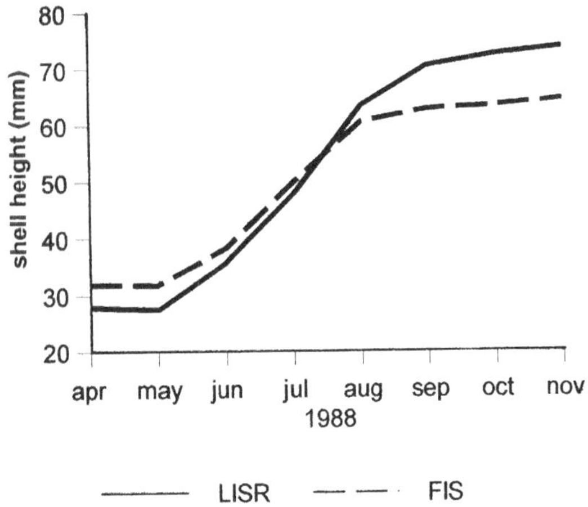

Fig. 9.2 Comparative rates of increase in shell height between resistant (LISR) and susceptible (FIS) oysters in Cotuit Bay, Massachusetts.

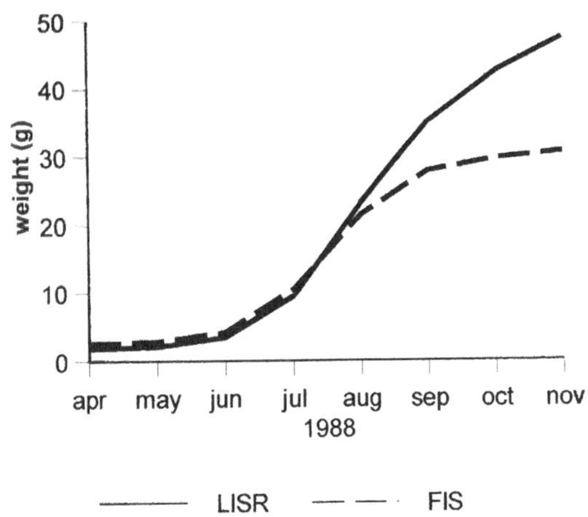

Fig. 9.3 Comparative rates of increase in whole weight between resistant (LISR) and susceptible (FIS) oysters in Cotuit Bay, Massachusetts.

percentage would have qualified for market, even though less than 2 years old.

Although the comparatively slow growth of the native oysters from Island Pond may have been due in part to MSX infection, it may also have been the result of genetic differences. The second phase of this investigation concerning possible approaches to the shellfish disease problem, funded in part by the National Science Foundation and conceived and implemented by Jonathan Davis, was the initiation of an oyster breeding program, with the intent of developing a strain of oyster that combined the dual advantages of rapid growth and disease (MSX) resistance.

In the spring of 1989, Ocean Pond Corporation obtained two groups of broodstock from experimental populations of oysters that had been tested for resistance to MSX in Cotuit Bay the preceding year. These were members of the sixth and seventh generations, respectively, of the Long Island Sound strain referred to above (designated as LISR) and a strain of oysters bred for resistance in Delaware Bay[9.6] (designated as CXF). The oysters in both groups had survived much of the impact of MSX in Cotuit Bay and had shown superior growth rate in comparison with their siblings. These oysters were conditioned and induced to spawn at the Ocean Pond hatchery in June 1989. The objective was to obtain embryos from 20 different pairs of parents, resulting in a series of distinct groups, or families, of known parental origin. If members of the different families showed distinct favorable qualities, such as superior growth rate, they would be saved for future broodstock.

Each member of the two groups of broodstock was identified by number. Spawning was induced in a shallow water table filled with slowly recirculating pond water warmed to the desired temperature by means of electric immersion heaters. As soon as each oyster began to discharge gametes, it was quickly removed from the spawning table, rinsed thoroughly in clean water, and then isolated in a plastic container and allowed to continue to spawn.

Eggs were obtained from four different females from the CXF group and one female from the LISR group. Sperm were isolated from four different males from both groups. Four separate batches of eggs from each of four different females were fertilized from four different males in a factorial arrangement, resulting in 16 groups of embryos having different parental pairs. In addition, four different groups of eggs from a single LISR female were fertilized with the sperm from four different CXF males.

The 20 groups of embryos were transferred to individual containers, where they were held until ready to set. They were then placed in individual setting trays containing small particles of dolomite as cultch. After setting had been completed, the cultch with spat attached was transferred to separate fly screen trays placed in the pond. By early September, all of the juvenile oysters were transferred to pearl nets suspended from buoyed long lines in the pond, where they remained for the winter. Unfortunately, four of the 20 groups were lost during harvesting in early spring.

In May 1990, samples of oysters were randomly drawn from each of the remaining 16 groups, placed in individual pearl nets, and transferred to long lines in Fishers Island Sound, where the relatively high salinity (30‰) and tidal currents provided a more typical oyster culture environment than the pond. At approximately monthly intervals, random samples were taken from each group, individually measured for shell height and weighed.

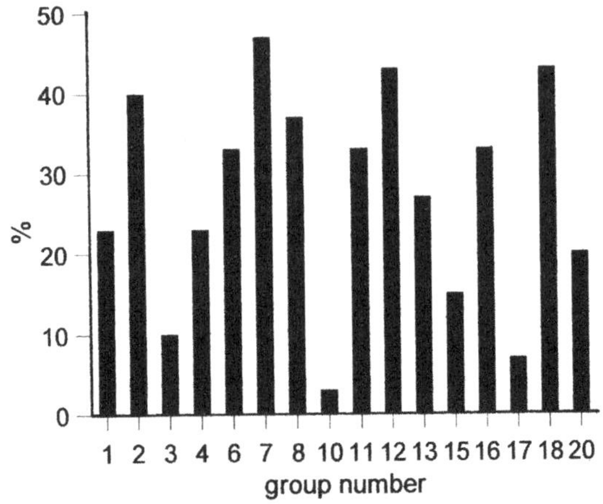

Fig. 9.4 Percentage of oysters, in different family groups, >75 mm in shell height and >50 g in whole weight, October, 1993.

The results of the final measurements of the 16 groups in October 1990, are shown in Fig. 9.4. It is evident that the progeny of certain of the parental combinations grew significantly faster than did those of other groups, even though all of the original parents were initially selected on the basis of size. This suggested that the usual procedure of inducing mass spawnings, with the eggs and sperm of many different males and females being randomly combined, could very likely produce poorer results than the mating of selected individuals whose growth characteristics had been established. The broodstock used at Island Pond today are of the same line as those that grew the fastest in this particular experiment.

The third phase of this investigation focused upon the possible advantages of culturing triploid eastern oysters in this region. Experiments in the induction of polyploidy in the Pacific oyster (*C. gigas*) had produced favorable results with respect to condition indices, and certain commercial hatcheries on the West Coast were concentrating on the production of triploid oysters[9.7]. Also, significantly faster growth among triploid *C. virginica* relative to their diploid siblings had been reported[9.8]. A rapid growth rate would presumably confer a distinct advantage on oysters exposed to possible MSX infection that were approaching marketable size. Furthermore, since triploid oysters expend relatively little energy on production of eggs and sperm, it was thought that their resistance to MSX infection during the summer months might be greater than that of diploid oysters that had spawned and whose reserves were depleted.

The procedure for inducing triploidy involved altering the number of chromosomes of embryonic oysters by exposure to the antibiotic cytochalasin B[9.8], and was undertaken by Jonathan Davis. The growth and survival rates of these oysters and their diploid siblings were compared over the course of two growing seasons at three different sites: Island Pond, Fishers Island Sound and Cotuit Bay. The growth rate of the triploids was found to be considerably greater than that for the diploids at all three sites[9.9]. Similar results were reported for triploid and diploid *C. virginica* grown in the York River, Virginia[9.10]. However, with respect to disease resistance, the potential advantages of polyploidy were not clearly established, even though the triploid oysters on the Cotuit beds experienced a slightly lower mortality rate than the diploid controls.

The diploid controls derived from oysters of the same selected group of parents as the triploids, but they were not necessarily true siblings, since they were obtained from a separate spawning. As was shown in the experiment with different family groups, oysters deriving from different parental combinations may exhibit distinctly different rates of growth, and this may have had an influence upon the comparative rates of growth of the triploids and diploids. It is also open to question whether or not different growth rates observed in the pond from year to year, or among different cohorts spawned in the same year, have been the result of varying environmental factors, such as food availability, or simply the result of slightly different genetic composition among the parents.

The major hope, as well as strategy, of most oyster growers today is to produce oysters that become of marketable size before they succumb to disease. One of the most useful results of the experiments described above has been that many of the oysters now produced by Ocean Pond become marketable when only 18 months of age. This is a significant improvement over previous years, when a minimum marketing age of 30 months, involving an extra year of disease exposure, was normally expected for the wild pond oysters.

10 Oyster Culture in North America

Until about 1950, much the largest percentage of world oyster production occurred in North America, primarily in the United States and, to a much smaller extent, in Canada. In 1913, the United States accounted for 88% of total world production, even though oyster landings in that year were considerably below those before the turn of the century[10.1]. The abundance of sheltered bays and fertile estuaries along its Atlantic and Gulf coasts provided a particularly favorable environment for the eastern oyster (*C. virginica*), the natural range of which extended from the Gulf of St Lawrence to the Yucatan Peninsula in Mexico. It is clear from historical evidence, in the form of extensive shell heaps, or middens, that this species was an important part of the human diet long before the arrival of the white man.

The other species native to North America, the Olympia oyster (*O. lurida*), is found from Alaska to the southern tip of the Baja Peninsula in Mexico[10.2]. Although prized as a delicacy in the American north-west, this oyster is comparatively small and, largely as a result of environmental degradation, quite rare. Today its contribution to total annual landings is less than 1%.

Two other species of commercial significance in North America, the Pacific oyster (*C. gigas*) and European flat oyster (*O. edulis*), were both deliberately introduced to the continent to supplement depleted stocks of native oysters. The former was imported from Japan to the Pacific coast early in the 1900s[10.3], where it adapted successfully and currently supports an important fishery in both the United States and Canada[10.4]. The European flat oyster was first introduced to the United States from France in 1949[10.5]. Although scarce, this species has established small localized populations in parts of Maine. Like the Olympia oyster, its contribution to total landings is very small.

In 1995, the United States was the second largest producer of oysters in the world. Landings that year approximated 239 000 t (whole weight), putting it slightly ahead of Japan and Korea but well behind China (Fig. 10.1). Approximately 85% of the landings were represented by the eastern oyster. Pacific oysters accounted for about 15%, the two species of *Ostrea* for less than 1% combined. In 1960, the United States ranked number one in world production and produced about 60% of the world total (discounting China, for which figures are not available). In 1997, production had fallen to about 211 000 t, ranking it behind both Japan and Korea and well behind China.

Most eastern oysters harvested commercially, perhaps as much as

95%, are from natural populations rather than hatchery-reared stocks. Historically, the major source has been from the natural beds in the Chesapeake Bay. In 1940, nearly half the total production of eastern oysters came from two states, Maryland and Virginia. Of lesser significance were the oyster populations in the Gulf of Mexico (primarily Louisiana)[10.6], the Middle Atlantic states (New York, New Jersey and Delaware)[10.7], and New England (primarily Connecticut)[10.8]. In New England, production declined sharply during the 1960s and 1970s, only to undergo a major revival in recent years (Fig. 10.2). However,

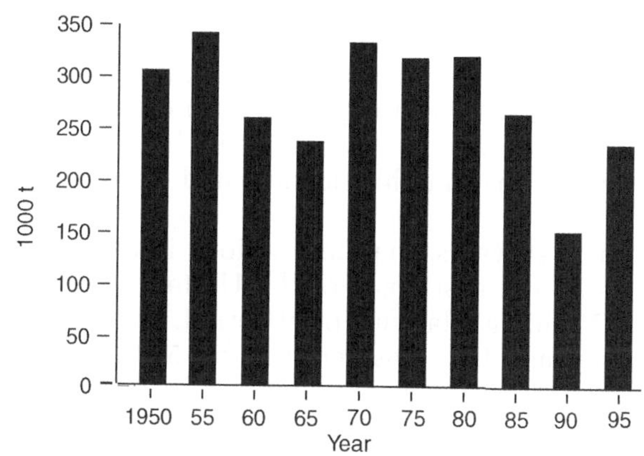

Fig. 10.1 Oyster production (whole weight) in the United States.

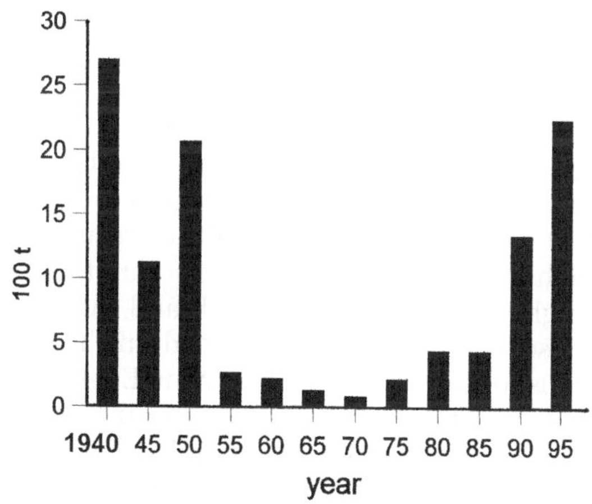

Fig. 10.2 Oyster production (meat weight) in New England.

the oyster industries in the Middle Atlantic states and Chesapeake Bay have experienced a severe decline, bordering on collapse, during the latter half of this century (Figs. 10.3, 10.4). (The 1995 harvest from the Chesapeake Bay was less than 2% of what it was in 1880). During the past 50 years, oyster production in the Gulf of Mexico has remained relatively stable (Fig. 10.5) compared with production along the Atlantic Coast, as has production of the Pacific oyster on the Pacific coast (Fig. 10.6).

The reasons for the overall decline of oyster resources in the United States vary in nature and degree from region to region and include a host of social, economic and environmental factors. Certainly an initial major cause common to all regions was lack of appropriate restrictive legislation and conservation management, resulting in the over-exploitation of what appeared to be a seemingly limitless but highly accessible resource, even before the turn of the century. The unhappy consequences of over-exploitation and mismanagement have been aggravated by a variety of factors, including water pollution, disease, coastal storms and hurricanes, and destruction of natural habitat.

During the past 40 years, the most serious impact upon the oyster populations along the east coast of the United States has been epizootic disease. Within a relatively brief period, 1958–1963, the harvest from the Chesapeake Bay was reduced by 50% as a result of the parasitic protozoan *H. nelsoni* (commonly referred to as MSX). Losses to the same disease were even more severe in the states of New Jersey and Delaware, where production fell by nearly 90% between 1957 and 1960. This and another pathogen, *Perkinsus marinus* (or 'Dermo'), are now endemic to these regions, and the recovery of the oyster populations is uncertain. Losses of this magnitude seriously reduce the parent stock and therefore reduce the likelihood of successful reproduction. Other factors possibly contributing to a poor recovery are deterioration in water quality and the sublethal effects of chemicals that adversely effect, in ways not clearly understood, the development of oyster larvae and juveniles.

During the past 50 years, oyster production in the Gulf States – Florida, Louisiana, Mississippi and Texas – has been quite variable from year to year. Sharp declines in this region are soon followed by rapid recovery. Despite significant mortalities resulting from *Perkinsus* and periodic hurricanes, natural oyster populations in these warm waters have a high reproductive potential, and fluctuations in abundance are not as severe as those experienced on the Atlantic coast.

In the late 1950s, oyster production in New England plunged to an all-time low, as successive years of set failure in Long Island Sound, New England's major producing area, combined with the destruction of oyster beds by a major hurricane, threatened to put an end to the industry. Since the late 1970s, however, the industry has begun a remarkable recovery. Prior to the recovery period, the water quality in

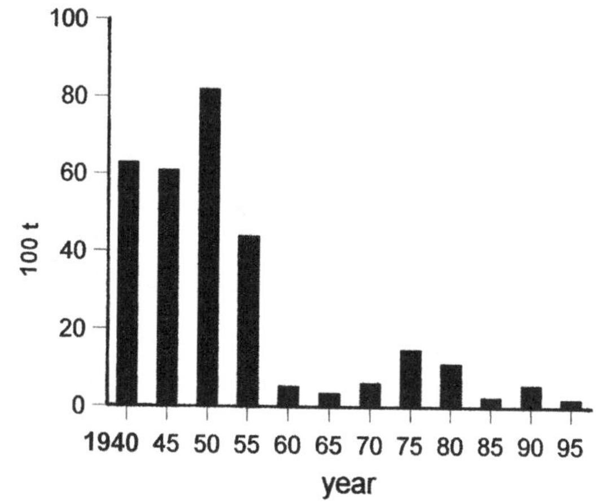

Fig. 10.3 Oyster production (meat weight) in Middle Atlantic.

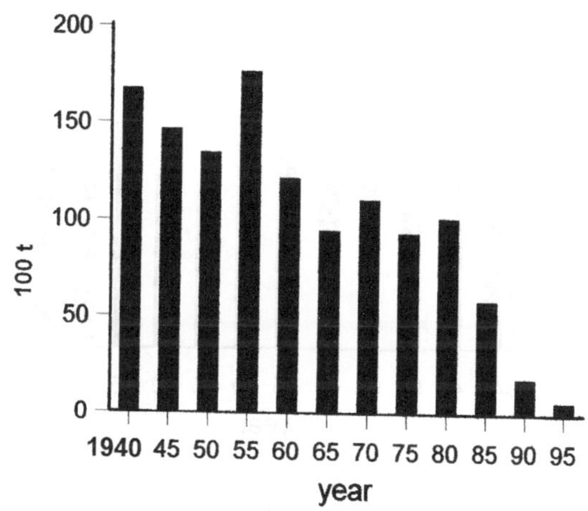

Fig. 10.4 Oyster production (meat weight) in Chesapeake Bay.

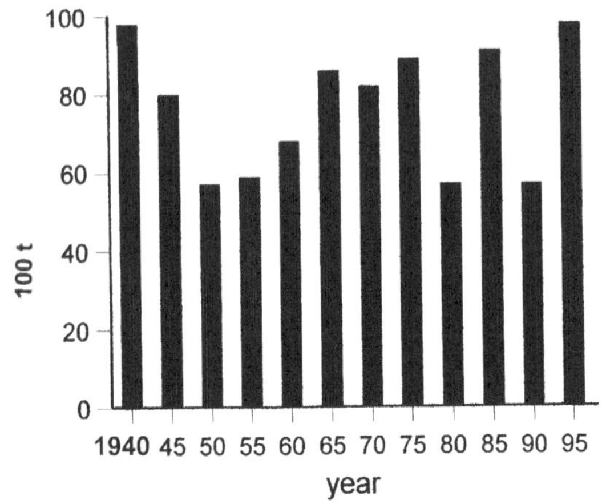

Fig. 10.5 Oyster production (meat weight) in the Gulf of Mexico.

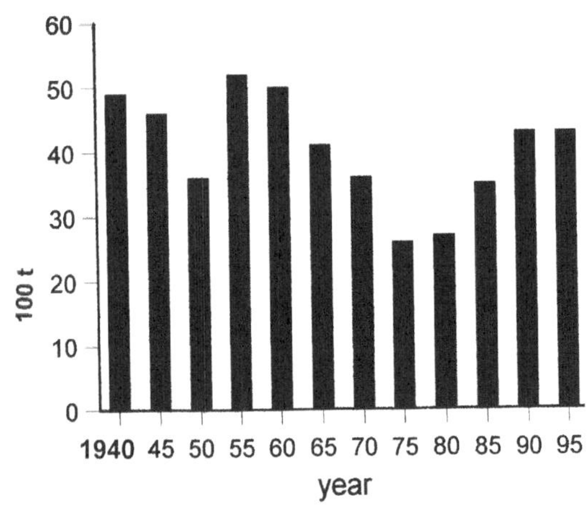

Fig. 10.6 Oyster production (meat weight) on the Pacific Coast.

Long Island Sound was greatly improved, and this, combined with an aggressive management program undertaken by the State of Connecticut in cooperation with a large private oyster company may have been chiefly responsible[10.9]. (This company controls nearly 8000 ha of bottom and dominates the New England oyster industry). Unfortunately, the recent occurrence of the two oyster pathogens, *Haplosporidium* and *Perkinsus*, in the Sound has been cause for considerable concern. Between 1995 and 1998, production dropped by nearly 80% as a result of heavy mortalities on the oyster grounds.

Methods of culture in the United States vary widely in terms of intensity and technical sophistication, and the degree of husbandry is generally a reflection of the potential market value of the product. Oysters that command the highest price are those destined to be sold live, in the shell, to be opened and served on the half-shell. A pleasing shape and appearance are nearly as important as flavor. The market price for oysters grown in New England is usually higher than the amount paid for oysters from more southern waters, where the combination of warm temperatures, low salinities and overcrowding is less conducive to quality than the colder, more saline waters of New England. Many of the culture techniques that have evolved in the United States, particularly during the past four decades, are in response to an expanding market for luxury food items, which include shrimp, lobsters and clams as well as oysters of superior quality.

Certain of the culture techniques employed throughout the industry were developed in Long Island Sound. Although this region was once well populated with oysters, which had been an important food for native Indians, heavy exploitation and overfishing had so diminished the supply by the end of the eighteenth century that oyster growers began to import oysters from as far south as the Chesapeake Bay. In 1858, 250 schooners imported 2 000 000 bushels to the New Haven area of Connecticut[10.7], a volume roughly equivalent to the total annual production from the Chesapeake Bay today. By the early nineteenth century, the practice of returning oyster shell to the bottom of bays and coves was initiated in an effort to improve recruitment. Oyster production in the New England region reached a peak of over 22 000 000 kg of shucked meat in the early 1900s.

The subsequent decline in the oyster resources of New England, with a concurrent rise in market price, stimulated efforts to develop systems of culture that would help assure a consistent supply of seed and, in effect, prevent the complete collapse of the industry. Of critical importance to the survival of the industry have been more intensive efforts to improve natural setting and growing areas and to reduce the abundance of predators, supplemented by the development of hatchery techniques for the production of juvenile oysters under controlled conditions. Many of the basic principles of larval culture worked out at the federal research laboratory in Milford, Connecticut[10.10,10.11], have been adopted by private growers and modified to suit their own needs. Although much the greatest percentage of national oyster production still derives from natural reproduction, some of the major producers, particularly on the Pacific coast now rely heavily upon hatchery production as their primary source of seed.

Commercial production of the eastern oyster in hatcheries is largely confined to the New England states, and the procedures at different hatcheries are basically similar. A prerequisite for all successful hatchery operations is the ability to culture algal food to satisfy the nutritional needs of the adults being conditioned for spawning, the larvae, and the post-larval juveniles. Although there have been many attempts over the years to develop an artificial food that might eliminate the need for maintaining algal cultures, none of these foods can supply a satisfactory diet without algal supplements. There are perhaps ten or more different species of microscopic single-cell algae that are currently used in hatcheries, and most are species that are small, have relatively thin cell walls and therefore are readily assimilated. Recently it has become possible to obtain certain species of algae that have been concentrated into a slurry and preserved. The slurry may be stored and/or shipped to other hatcheries for future use as a supplementary food. Most hatcheries, however, have the capability of maintaining their own algal cultures.

In areas where natural reproduction is inconsistent or infrequent, hatcheries can provide an alternative or at least supplementary source of seed oysters. They also offer an important tool for selective breeding, enabling the choice of adult broodstock to develop lines genetically more resistant to disease or having superior growth attributes, or genetic manipulation of ploidy for similar purposes. Nevertheless, the great majority of eastern oysters produced in the United States today derive from natural reproduction. Hatchery production involves additional costs and is not devoid of problems, such as water contamination and disease infections, and one heavy natural set of oysters often provides several years of satisfactory production.

In major producing areas such as Louisiana, Connecticut and the Chesapeake Bay, natural reproduction is assisted by planting large quantities of shell as cultch each year in anticipation of obtaining sets from the wild oyster populations. The larger oyster producers in Long Island Sound may lease many thousands of hectares of bottom for private culture purposes. Most of the natural seed beds are located along the Connecticut (north) shore of the Sound, often near river mouths, and frequently in the rivers' lower reaches. Here the water temperatures may become appreciably higher than in the Sound itself, and it is believed that these warmer estuarine areas are the primary source of oyster larvae for the region.

The private oyster growers in Connecticut obtain their seed either from independent fishermen, who harvest from public beds, or from their own seed beds. In order to obtain a natural set, empty oyster shells are collected from shucking houses or dredged from the bottom and stored on shore to air-dry, preferably for several months (Fig. 10.7). This exposure helps remove the remains of fouling organisms, leaving the shell relatively clean. When it is believed on the basis of larval samples, water temperature or intuition that spawning has occurred, the shell is transported to the setting beds by boat and spread over the bottom in amounts up to 2000 bushels (about 70 m^3) of shell per ha (Fig. 10.8). Prior to spreading the cultch, however, the beds are worked over with suction dredges to remove predators, such as oyster drills and starfish, and old shell that is fouled.

In Long Island Sound, setting occurs in mid- to late summer. The seed is usually allowed to remain on the setting beds through the remainder of the fall and winter. During this period, the tiny oysters gradually become covered with a fine layer of silt. Since these would smother when temperatures rise and metabolic activity increases in the spring, the shell with seed attached is dredged up in late winter and transported to the growing grounds. Here tidal currents are generally swifter than on the seed beds, and oyster growth is faster. The shell with seed is spread on these beds at an average density of about 1000 bushels (35 m^3) per ha.

Unfortunately for oysters growing on the bottom, their life is a precarious one. In addition to siltation, there is the ever-present threat of natural predators, of which there are a variety. Perhaps the most destructive oyster predators in Long Island Sound are the starfish (*Asterias forbesi*) and the oyster drill (*Urosalpinx cinerea* and *Eupleura caudata*), and a great deal of thought and energy has been given to their control[10.9]. Because of their spines, starfish can be entangled in large mops dragged over the bottom, brought to the surface, and immersed in tanks of hot water on the boat deck. Granulated quicklime, to which the epidermis of starfish is quite sensitive, may also be spread over the beds. As mentioned earlier, populations of oyster drills can be controlled to some extent by means of suction dredges that vacuum the drills up from the bottom. These methods have proven effective, but they represent a significant investment in labor and equipment. One company owns and operates over 20 vessels that clean the beds, spread the shell, and dredge oysters for transplanting or processing for market.

Because of the relatively low water temperatures in Long Island Sound, oysters generally do not become marketable before 3–4 years of age. Before being harvested for market, however, they are sometimes shifted in the spring for a final summer of growth on fattening beds, where the meats improve in quality. The majority of oysters harvested from Long Island Sound are destined for the half-shell market, where value is based upon size and appearance as well as taste.

The methods for oyster production in the Middle Atlantic States and Chesapeake Bay are similar to those described for Long Island Sound in that the oysters are grown on, and harvested from, the bottom, in subtidal areas, and the oyster populations rely for the most

Fig. 10.7 Shell being stockpiled for cultch, Long Island Sound (Connecticut Department of Aquaculture, courtesy of John Volk).

Fig. 10.8 Shell being loaded on oyster boat, Long Island Sound (Connecticut Department of Aquaculture, courtesy of John Volk).

part on natural reproduction for recruitment. In New Jersey, Delaware, and Virginia, natural seed beds traditionally have been developed and maintained by state agencies, largely through provision of cultch and some degree of predator control. Fishermen with private leases of their own harvest the seed from the public beds and plant these on their own grounds. Since the late 1950s, when the natural beds in the Delaware River and lower Chesapeake Bay were almost eliminated by disease, much effort has been devoted by state research facilities in this region to developing lines of disease-resistant oysters. Whether or not this approach will eventually succeed in restoring the beds and reviving the industry is unclear, since many generations of selective breeding appears to be required in order to achieve moderate improvement in survival.

The oyster industries in Maryland and Virginia are structured quite differently in that the Maryland oyster fishery is almost entirely public, and private control of bottom grounds is discouraged, whereas about half of the oyster-producing areas in Virginia waters are privately leased. Currently the State of Maryland is undertaking a major effort towards restoring the natural oyster bars through intensive planting of shell on public grounds, but this project is jeopardized by the dual threat of the parasites *Haplosporidium* and *Perkinsus*.

Oysters in the Chesapeake Bay require 2–3 years to reach marketable size. Owing to the shallowness of the water in much of the bay, many of the oysters are harvested by means of hand tongs (Fig. 10.9). In deeper water, dredges (Fig. 10.10) and patent tongs (Fig. 10.11) are generally employed. In recent years, some of the fishermen have begun to use diving gear, picking select oysters off the bottom by hand.

Oyster production in the Gulf of Mexico presently exceeds that of the Chesapeake Bay, once by far the leading producing area. The number one state in this region as well as in the United States is Louisiana, where the industry has remained relatively stable during the past 60 years[10.12]; landings actually increased during the 1980s while those along the Atlantic coast were in a sharp decline. In recent years, annual production has approximated 4 000 000–5 000 000 kg of shucked meats, which exceeded the entire production from Maryland and Virginia combined. Although roughly one-quarter of the growing grounds in Louisiana (over 500 000 ha) are privately leased, the industry and its methods have changed very little since motorized vessels replaced sail in the 1920s.

As in the case of most of the oystermen on Long Island Sound, Delaware River and Chesapeake Bay, Louisiana growers rely entirely upon natural reproduction as the source of supply of seed. The state of Louisiana assumes responsibility for the planting of cultch on public seed grounds each year. The private growers transplant the seed from the seed grounds to their private bedding grounds.

The preferred cultch in Louisiana is clam (*Rangea cuneata*) shell. The shell is spread on the seed beds in the spring, as spawning in

Fig. 10.9 Harvesting oysters by tongs, Chesapeake Bay (courtesy of Chesapeake Bay Maritime Museum).

these warm latitudes may occur in early May. Certain growers maintain their own seed collecting grounds, which are located near the mouth of the Mississippi River. The young oysters are relatively safe on the seed beds, since neither their major predators, primarily the carnivorous snail *Thais haemostoma* and certain species of finfish, nor the pathogen *Perkinsus marinus* are tolerant of low salinities.

The seed usually are not transported to the high salinity growing grounds in the Gulf until 3 years of age. Here they grow rapidly and improve in quality and flavor in a few months time. If allowed to remain on these beds during the warm summer months, however, the chances of survival are reduced owing to the likelihood of infection by *Perkinsus*. The oysters of superior quality are sold for half-shell consumption. The majority, however, are steamed open and canned for distribution.

Despite the fact that off-bottom culture of oysters is practiced quite successfully in many areas of the world, e.g. Korea, China and Japan,

Fig. 10.10 Harvesting oysters by dredge, Chesapeake Bay (courtesy of Chesapeake Bay Maritime Museum, photo by Robert De Gast).

Fig. 10.11 Harvesting oysters by patent tongs, Chesapeake Bay (courtesy of Chesapeake Bay Maritime Museum).

production of the eastern oyster is limited for the most part to subtidal bottom culture. Several attempts at raft culture of this species, using the Japanese technique of hanging strings of shell bearing the seed, have been attempted without great success. Although growth is rapid and survival high, the shells of the oysters may be too thin and fragile to be fully satisfactory for the half-shell market[10.13]. Furthermore, rafts are vulnerable to ice, heavy winds and boat traffic[10.14], and it is difficult to protect the oysters from theft. Finally, the process of assembling strings of shell and constructing rafts is initially far more expensive than simply spreading the oysters on the bottom. However, nets, trays and racks are used successfully in certain areas by relatively small producers not requiring a large amount of growing area.

A huge amount of research and effort has been devoted by federal, state, academic and private agencies, institutions and individuals during the past 40 years to various strategies and methods for arresting the decline of the eastern oyster. The major efforts have concentrated where the need was most critical, as in the New England states where production in many areas had all but ceased, and in the Middle Atlantic and Chesapeake Bay regions, where, in some areas, disease has threatened to eliminate the entire industry[10.15]. This research has led to a great deal of published information on the feeding and rearing of oyster larvae under controlled conditions, inducement of setting on a variety of substrates, raising of the juveniles under controlled or semi-controlled conditions, predator control, identification of toxic materials in the environment, oyster genetics, oyster diseases and other topics. There is no doubt that this effort has helped sustain an industry that, in certain areas at least, might otherwise have collapsed. Furthermore, although much of the research has been concerned with the Middle and North Atlantic regions of the United States, and with the native oyster *C. virginica*, it has nevertheless been of broad application and of distinct value to shellfish culturists the world over. However, the fact remains that, during this 40 year period, national production of oysters has not increased but has in fact significantly decreased, and the feasibility of introducing the Pacific oyster to parts of the Atlantic coast has been seriously considered[10.16,10.17].

In 1995, the Pacific oyster accounted for about 15% of total United States oyster production. This industry is based upon private control of oyster beds, providing a strong incentive for husbandry[10.4]. Also, natural reproduction of this species is limited in this region, and seed oysters from Japan, once imported in large quantities as the foundation of the industry, are no longer available at a price or of a quality acceptable to many Pacific coast oyster growers. It became necessary, therefore, to turn to hatcheries as a major source of seed, and certain of these have developed into significant producers.

The Pacific oyster was first successfully introduced to the state of Washington in the early 1900s in an effort to sustain an industry historically dependent upon the native, or Olympia, oyster (*O. lurida*).

Between 1890 and 1915, production of the latter species had dropped by over 90% as a result of over-exploitation, lack of management, and adverse weather conditions, and efforts to initiate an industry based upon the American oyster, imported from the east coast, had failed. Shipments of the Pacific oyster in commercial volume from Japan were begun in 1921. Except for a very few locations, such as Dabob Bay in Washington, summer water temperatures along the United States' north-west coastline are too low for this species to reproduce with any regularity. Therefore, the industry relied heavily upon the imported seed until the 1970s, when the development of hatcheries in the Pacific north-west resulted in a less expensive and more dependable alternative. After reaching a low in 1975, production increased rapidly during the next 15 years and has since levelled off to around 10 000 000 lb (4 500 000 kg) of shucked meat (or around 37 000 t, whole weight) per year (Fig. 10.6).

The traditional method of growing *C. gigas* in the Pacific north-west is by bottom culture. Off-bottom culture techniques are more costly and often not justified by market price. The seed, whether imported from Japan, collected from regional spat-collecting areas, or obtained from hatcheries, are usually attached to whole oyster shells which are spread on the bottom. At most hatcheries, plastic mesh bags are filled with oyster shell and immersed in tanks containing late-stage larvae (Fig. 10.12). After setting is completed, the bags are removed from the

Fig. 10.12 Setting tanks, with mesh bags of oyster shell as cultch, Pacific Coast (photo by O. H. Matthiessen).

Fig. 10.13 Moving bags of cultch, with spat attached, to growing grounds, Pacific Coast (photo by J. P. Davis).

tanks and transferred to the growing grounds (Fig. 10.13), where the shell, with spat attached, is spread on the bottom. Because of large tidal amplitudes, as much as 6 m in certain areas, expansive areas of bottom are exposed during low tides, and most of the oyster beds are intertidal. In areas where the bottom is soft mud, shell or gravel is spread to provide a firmer substrate for the seed during its first year of growth.

As the young oysters grow, the clusters are broken apart and spread over the growing grounds, the planting density being reduced from around 8 000 000 seed per ha to perhaps 200 000 oysters per ha. In some areas, the oysters are worked over, and ultimately harvested, by hand. The larger growers use bag dredges or escalator dredges over the beds at high tide in order to shift oysters from one bed to another, break apart the clusters, and to harvest. It requires an average of 3–4 years after planting for the oysters to reach a satisfactory size for processing. The majority of oysters produced on the Pacific coast are shucked and the meats marketed either fresh, frozen or canned.

Off-bottom culture is practiced in certain areas, often where suitable bottom is either absent or unavailable. Although off-bottom culture methods require a greater investment in gear and maintenance, certain of the problems inherent in bottom culture – predation by the

Fig. 10.14 Strings of oyster shell on intertidal racks, Pacific Coast (photo by J. P. Davis).

Japanese oyster drill (*Ocenebra japonica*) and Dungeness crab (*Cancer magister*), and smothering of the seed oysters by the feeding and burrowing activities of shrimp – are avoided. Trays containing oysters or strings of shell bearing the seed may also be suspended from rafts. On some beds, shells with spat attached are suspended in strings from racks positioned low in the intertidal zone (Fig. 10.14), or juvenile oysters obtained from hatcheries as 'singles' are held in plastic mesh bags supported off the bottom on racks made of iron rods, as done in France (Fig. 10.15).

Production and shipment of eyed larvae for 'remote setting' has become a growing industry on the Pacific coast in recent years[10.18]. Larvae that have reached setting size are filtered from hatchery tanks, using a sieve of appropriate mesh size, and wrapped in a damp cloth. They are then placed in a styrofoam box and shipped to the buyer, who transfers them to tanks filled with warm sea water and algal food and containing the cultch material, usually bags or strings of clean oyster or scallop shell. As a rule, the producer of the larvae guarantees a certain percentage survival through setting and may also supply stored algal slurry for food if requested by the grower. If held at temperatures near freezing, the larvae may survive in the shipping container for up to 7 days. This system has proven useful since oysters still in the larval stage are considerably less expensive than small

Fig. 10.15 Intertidal racks supporting mesh bags of oysters, Pacific Coast (photo by J. P. Davis).

seed, and the chance of disease transmission is considered to be much lower.

A growing number of producers of Pacific oysters sell their product alive, in the shell, directly to restaurants[10.19]. These oysters are usually hatchery-produced and are cultured off-bottom, with considerable attention given to size, shape and appearance. The cultch used for collecting the spat consists of fragments of oyster shell, fine enough to ensure a high percentage of single oysters. Once attached, the spat are transferred to upwellers, to which food is added. When the oysters reach about 10 mm in size, they are removed from the upwellers and placed in wooden trays with plastic mesh bottoms. The trays are stacked in tiers and hung from long-lines in a nearby bay. At a size of 25 mm or more and about 4 months old, the oysters are transferred to nets suspended from long-lines, or to plastic mesh bags on racks, where they remain until harvested.

A major problem afflicting the Pacific oyster industry is summer mortality, a phenomenon that apparently also occurs in Japan. These mortalities usually occur in growing areas of high natural productivity and high nutrient level, and at temperatures in excess of 20°C, and coincide with the period of maximum gonad development. Mortalities may reach 50% in some areas. It has been suggested that these oysters, stimulated by high food concentrations and temperatures favorable for the development of gametes, have expended their stored reserves and may be particularly susceptible to environmental stress, such as heat exposure at low tide[10.20], although the bacterium *Nocardia* has been implicated as well[10.16]. This problem has been alleviated to some extent by the inducement of triploidy at fertilization. Triploid oysters are essentially neuter and, unlike diploids, do not channel large amounts of energy into the development of gonad material but rather into growth.

Returning to the east coast of the United States, the European flat oyster (*O. edulis*) has succeeded in establishing small isolated populations along the coast of Maine[10.21]. Because this species thrives in cold and clear water of high salinity and spawns readily at temperatures as low as 16°C, it is well adapted to the waters north of Cape Cod. However, it is not sufficiently abundant to support a stable fishery. Perhaps because of its scarcity, the flat oyster often commands a higher market price than any of the four species produced in the United States. It is very much a gourmet item, traditionally served raw on the half-shell.

The majority of flat oysters harvested commercially in Maine derive from wild stocks. However, small numbers have been cultured, using off-bottom techniques (K. Tacy, pers. comm.). The seed are obtained from hatcheries at a size of 2–3 mm and held in floating trays during their first growing season. The following season, they are held in lantern nets suspended from surface long-lines, at the end of which they may have reached up to 65 mm in size. The oysters are placed on the bottom during their third season to improve the shape and hardness of shell, and those in excess of 55 g in weight by fall are marketed.

The native oyster of the Pacific coast, the Olympia oyster (*O. lurida*), is also of relatively minor importance[10.22]. This is a slow-growing species, requiring up to 4 years to reach a commercial size of about 50 mm. It is sensitive to both heat and cold, and the system that evolved for its culture required construction of dikes around the intertidal growing beds in order to regulate the water level and keep the oysters submerged as necessary. Shell is frequently scattered on the beds to collect spat from natural reproduction. However, there is little effort to culture this species off-bottom, since fouling by mussels and barnacles in its chief area of abundance (Puget Sound) is intensive. Within the diked areas, the oysters are thinned and predators removed by hand. When harvested for market, the oysters are shucked and sold by the gallon.

Apparently this species was particularly vulnerable to overexploitation because water temperatures in Puget Sound were not warm enough to assure successful reproduction each year. Overharvesting of the adult population, therefore, quickly led to a decline in abundance. Attempts have been made to supplement natural sets with hatchery-produced seed, but the majority of growers find the hardy and fast-growing Pacific oyster to be a more dependable crop.

Furthermore, maintenance of the dikes necessary for the protection of the oysters add appreciably to the cost of production.

The oyster industry of Canada is small but relatively stable, averaging between 5000 and 6000 t during the past 5 years. The industry is based upon two species, the eastern oyster on the Atlantic coast, and the Pacific oyster on the Pacific coast. In 1995, about 55% of the Canadian landings consisted of Pacific oysters.

Production in Canada is constrained by a relatively brief growing season due to low water temperatures[10.23]. Generally, a minimum of 5 years is required for the eastern oyster to attain a marketable size of 75 mm in the Maritime Provinces. Even on the Pacific coast, where ice is not a factor, 4–6 years are required for the relatively rapid-growing Pacific oyster to mature[10.22].

On Canada's Atlantic coast, most oyster production occurs in Malpeque Bay. It is largely a public fishery with few private leases, and the government manages the beds by spreading shell, transplanting oysters from unfavorable or contaminated areas to clean areas, and cultivating the bottom. Wild spat are collected in natural setting areas, using shell spread on the bottom or suspended in mesh bags from fences in shallow water[10.24]. Oysters relaid from contaminated onto clean growing areas contribute significantly to total production, in some years up to 80%[10.25]. Most of the oysters harvested, usually by hand or by tongs, are sold for the raw-oyster market.

Since the first commercial shipment of Pacific oysters arrived on Canada's west coast from Japan in 1925, oyster growers generally relied upon imported seed[10.26]. Since 1962, however, the majority of seed has been obtained from natural reproduction in particularly favorable setting areas, notably Pendrell Sound, and, more recently, from hatchery production. The great majority of Pacific oysters are grown on intertidal bottom in British Columbia. Most of the production occurs on relatively small family-operated leases, and the total area available for production is limited. The seed, usually collected on oyster shell, is spread on firm ground high in the intertidal zone. After 1 year, when large enough to survive in softer sediment, they are moved lower in the intertidal zone. The oysters are eventually harvested by hand rake, or by dredge at high tide.

Increasing interest in a half-shell market for local oysters has encouraged some growers in British Columbia to adopt off-bottom culture techniques[10.27]. In addition to improved yields per unit of culture area, other advantages to off-bottom culture would include improved quality of the oyster meats and the ability to culture in, or over, areas unsuitable for bottom culture. Growers interested in producing oysters for the half-shell market generally prefer to work with single oysters, cultured in trays or nets, rather than with oysters clustered on shell and suspended on strings.

Some of the private growers in British Columbia purchase eyed larvae from hatcheries and utilize the so-called 'remote setting' technique described earlier[10.18]. This process usually begins in the middle of April and continues through October. Plastic mesh bags of clean oyster shell are suspended in tanks filled with filtered and heated sea water. After the larvae are added to the tank, the bags remain in the tank for about 1 week before being removed. They are then suspended from plastic floats for about 1 month before the cultch with spat attached is spread in the intertidal zone.

In contrast with the Canadian Maritime Provinces on the Atlantic Coast, most of the oyster production in British Columbia comes from private leases.

11 Limits to Oyster Production

Oysters have been frequently cited as excellent candidates for culture in areas of the world where sources of protein food are scarce. Unlike finfish, crustaceans such as shrimp and lobsters and most other edible marine animals selected for culture, oysters are sedentary rather than fugitive and do not require enclosures. Because of their low position in the aquatic food chain, and because they are capable of obtaining all their necessary nutrition from the surrounding water, oysters do not require supplemental feeding as do most other species in captivity. Oysters are also remarkably hardy in certain respects, capable of withstanding a broad range in temperature and salinity as well as prolonged exposure to desiccation. Finally, oysters are considered to be exceptionally high in nutritional value, more than 50% protein, low in carbohydrates and fats, with a complete assortment of amino acids and high in vitamins and trace elements. Given these accomodating qualities, one might suppose that growing oysters would be a relatively easy proposition. Such is not the case.

The great majority of the oysters harvested today are from a relatively narrow band that circles the globe between latitudes 50°N and 20°N. It might be supposed that, at least partially in order to alleviate local human nutritional needs, great numbers of oysters would be cultured in tropical waters, with even temperatures year around, but this is not true. Between the Tropics of Cancer and Capricorn, only Thailand, the Philippines, and Mexico currently produce oysters in excess of 10000t per year. Combined landings in the huge continents of South America and Africa amount to less than this number, although it is likely that, in low population areas, oysters are harvested for direct local consumption and go unreported.

The natural limitations upon biological productivity in the marine environment are one reason for this. The fact that the great percentage of world oyster production occurs in north temperate latitudes is not simply the result of more advanced technological development or the demography and economic infrastructures of these regions. While production in polar regions is virtually non-existent because water temperatures are too cold, production in many parts of the tropics is limited as well, even though climatic conditions would seemingly be favorable for year-around growth. In certain areas, poor oyster production may be due to the fact that warm surface layers, in which the oysters might ordinarily grow, become depleted of nutrient material, and only vertical mixing or upwelling can bring these nutrients back to the surface where photosynthesis can occur. Photosynthetic production of carbon per unit area per day in much of the tropics may be less than one-tenth the amount produced in inshore temperate regions[11.1]. Although small amounts of oysters are grown in Jamaica, Cuba, and in other parts of the Caribbean, production is confined for the most part to mangrove lagoons and inlets receiving nutrient-rich drainage from the land, and, even in these areas, the oysters may be relatively small and have poor meat quality. The meat yield from a tropical oyster such as *Crassostrea rhizophorae,* harvested from the roots of a mangrove at a length of 2 inches, amounts to very little.

Low primary productivity, therefore, is one reason why large-scale culture of oysters in much of the tropics has not been particularly successful. In areas influenced by sufficient land runoff to provide the necessary inorganic nutrients and particulate organic matter necessary for food, there may be other natural factors that limit oyster production[11.2]. In estuarine areas, seasonal rains may cause an abrupt drop in salinity and sharp increase in turbidity of the coastal waters that, in severe cases, can result in heavy mortalities. While growth may occur year-around, so too does the incidence and growth of fouling organisms, which can smother the oysters and/or greatly reduce the efficiency of the culture gear. Included in the category of fouling organisms are juvenile oysters that, having settled in large numbers on the adults as larvae, attach and soon become serious competitors for space and food.

In those parts of the world where the environment is favorable for growth, the production of shellfish by suspension, thereby utilizing all three dimensions of the water column, may achieve a greater yield of animal protein per unit area than any other form of farming[11.3]. Perhaps the most outstanding example is sea mussel culture in the Galician Bays of Spain, where the mussels, attached to rope lines 10 m or more in length, are suspended from rafts. Somewhat lower, but none the less impressive, annual yields are obtained on oyster farms in parts of Japan and France[11.4], which are among the top five oyster producers in the world. Although the production systems in the two countries are quite different in certain respects, there are also certain similarities that help explain their ability to achieve exceptional levels of production.

In Japan, where hanging culture techniques are employed, production from rafts may be as high as 20 000 kg of shucked meat (wet weight) per ha per year, and, from long-lines, up to 26 000 kg/ha per year[11.4]. These estimates were based upon extrapolations from relatively small culture units, e.g. the surface area of individual rafts. In Hiroshima Bay, where more than 80% of Japan's oysters are produced, approximately 3500 ha are utilized for oyster production, and the annual sustainable yield is about 20 000 t[11.5]. On this basis, annual production per unit area is perhaps closer to 5000–6000 kg per ha per year. Attempts to obtain higher yields by increasing the number of rafts have been quickly followed by a marked decline in production, accompanied by deterioration in oyster quality[11.6]. This may be partially due to declining water quality as well as to the fact that the carrying capacity of the Bay is being exceeded.

Nearly comparable levels of production are obtained in some of the major growing areas of France, even though surface suspension methods are not employed and the majority of oysters are grown either on intertidal racks or directly on the bottom. Approximately 2000 m^2 of intertidal bottom reportedly can yield up to 30 t of oysters (whole weight) per year (E. His, pers. comm.), which would convert to about 22 500 kg of meat per hectare per year. Again, these particular yields were determined on the basis of small culture units and are not representative of extensive culture areas. Perhaps a more realistic estimate of annual productivity can be obtained from statistics for the Marennes-Oleron region, which produces about 40 000–60 000 t (whole weight) of oysters each year, the most for any area in France[11.7]. Since about 4000 ha are under cultivation, the annual yield per ha is about 1500 kg of meat. This is similar to typical yields obtained by growers in the Bay of Arcachon, where an annual harvest of 2000 kg meat per ha per year is not unusual.

Although their systems of culture differ, the oyster industries of Japan and France are similar in several respects. Both industries are now based largely upon the Pacific oyster (*C. gigas*), native to Japan but not cultured in France until the 1970s. The Pacific oyster is a hardy species, remarkably tolerant of the various diseases that have decimated populations of *C. virginica*, *C. angulata* and *O. edulis* in North America and Europe. Its rapid rate of growth combined with a high meat weight:total weight ratio accounts in large part for the high yields described above. Also, annual reproduction has been relatively dependable

Furthermore, the Inland Sea of Japan, which includes Hiroshima Bay, and the Marennes-Oleron district of France share certain geographic and hydrographic characteristics. Both are in north temperate regions and have a similar climate. In Hiroshima Bay, water temperatures rarely drop below 10°C, and the summer maximum is usually about 26°C. In the Bay of Oleron, the winter minimum is about 8°C, with a summer maximum of about 20°F. Under these temperature regimes, it is possible for the Pacific oyster to grow virtually year-around, a factor that contributes to these exceptionally high yields. (This species may feed at temperatures as low as 5°C, compared with a minimum of about 10°C for *C. virginica*[11.8]). Both areas are spared the occurrence of ice, which normally might preclude the types of culture gear used in both countries, and both receive the input from rivers, providing inorganic nutrients to enhance primary productivity and assure a plentiful amount of planktonic food in the oyster culture areas. The tidal amplitude in Hiroshima Bay is 3 m and, in Marennes-Oleron, up to 4 m, with strong currents occurring in both areas. All of these factors – minimum water temperatures of about 10°C, a constant source of nutrients from river flow, and strong tidal currents that continually replenish the oyster grounds with food – are conducive to the rapid growth of this species.

It is evident that, during the past 15–20 years, the growth curve for oyster production in both Japan and France has flattened. This may be a result not of too few oysters but of too many. In Hiroshima Bay, where up to 80% of Japan's oysters originate, a marked decrease in oyster growth rate became apparent by the 1970s, as the yield per raft dropped from 2.5t to 2.0t[11.6]. Meat yields began to drop from 30g per oyster in 1950 to 15–18g by 1980, and the time required for an oyster to reach market size increased from 6–12 months to 3 years[11.5]. This decline in growth rate was accompanied by higher mortality rates, which may have been related to deterioration in water quality as well as to overcrowding. Although the biologists of Hiroshima Prefecture would have preferred a maximum of 10000 rafts deployed in Hiroshima Bay, a limit of 12000 rafts was eventually established[11.9].

A similar situation developed in France, particularly in the Marennes-Oleron region, where the greatest percentage of France's oysters is produced. As the total oyster biomass in the Bay of Oleron rose to 200000t in the early 1980s as a result of successive years of intensive setting, growth rate began to drop. Oysters that once reached a size of 80g in 18 months would now require 4 years or more to attain this size (M. Heral, pers. comm.). A limit of 6000–7000 bags per ha, or stocking biomass of 72t/ha, was therefore imposed, with the objective of stabilizing annual production at a level of 40000–60000t[11.7]. In addition, an effort has been made to reduce the number of oyster competitors on the beds, such as the slipper shell *Crepidula fornicata*, by periodic dredging.

Although less is known about the immense oyster industry in China, there are indications that the carrying capacity of its waters, at least in certain areas, may be being exceeded as well[11.10]. Overcrowding in estuaries already taxed by industrial pollution and human wastes could well lead the way to epizootics and massive mortalities, and it may well be that oyster production from some of these major producing areas will not increase appreciably.

If the requirements for a rapid rate of production include a fast-growing species, moderate water temperatures year-around, constant tidal exchange and a source of nutrient-rich water as discussed above, a production rate in Island Pond similar or even close to that reported for Hiroshima Bay or the Marennes-Oleron region in France would certainly not be expected. As mentioned earlier, the American oyster is a relatively slow grower at water temperatures below 15°C in comparison with the Pacific oyster, and the meat weight: total weight ratio is smaller owing to its heavier shell. Because of winter water temperatures that may drop to freezing, there is no growth during at least one-third of the year. There is no source of nutrient-rich water flowing into the pond from the land, and tidal currents are negligible. The only water movement is wind-induced and therefore sporadic and limited for the most part to the surface. Under these conditions, a relatively low production rate would be expected.

In Island Pond in 1997, roughly 1 600 000 seed oysters spawned during the spring were cultured until fall in the south sector of the pond, an area comprising about 7.5 ha. On the basis of a series of samples collected during the fall, it is estimated that the total wet meat weight of this entire crop of oysters at this time was at least 2500 kg. Therefore the amount of production of cultured oysters for the season, not counting that for wild oysters in the same area, would exceed 300 kg/ha, or about twice the 150 kg/ha per year production rate predicted for an enclosed body of water[11.3]. During 1986, approximately 800 000 yearling oysters were cultured in the same area of the pond and added a total of about 4000 kg in wet meat weight prior to the end of the growing season, indicating an annual production of over 500 kg/ha. This is well below the levels of production cited above for France and Japan, and it is questionable whether production could be increased through more intensive culture practices.

The standing crop of phytoplankton in Island Pond during the growing season compares very favorably with that of the open waters of Fishers Island Sound. The total numbers of planktonic cells available as food may at times exceed that in the Sound by a factor of ten. This superiority in abundance of planktonic food is also reflected in comparative total chlorophyll measurements in the two areas, which have indicated the pond concentration to be nearly five times that occurring in the Sound. This, in combination with the higher water temperatures that prevail in the pond during the growing season, would most likely account for the discrepancy in growth rate between the two areas. Although rate of shell growth has been observed to be greater in the Sound, the ratio of meat weight to total weight has been found to be considerably greater among the pond oysters.

A series of chlorophyll measurements was recently made in the pond during the growing season (April–November) to determine whether or not the standing crop of food was being depleted significantly during the course of a growing season and therefore whether the carrying capacity of the pond was being exceeded. The chlorophyll concentration was observed to rise during the spring, coinciding with the harvest and removal of the 1 year old oysters from the pond, then reach a peak in mid-summer when the biomass of oysters was low, and decline during late summer and fall, as the new crop of oysters, spawned in the spring, was increasing rapidly in biomass. Although a relationship is suggested, it is obscured to an extent by the relationship between chlorophyll and water temperature (Fig. 11.1).

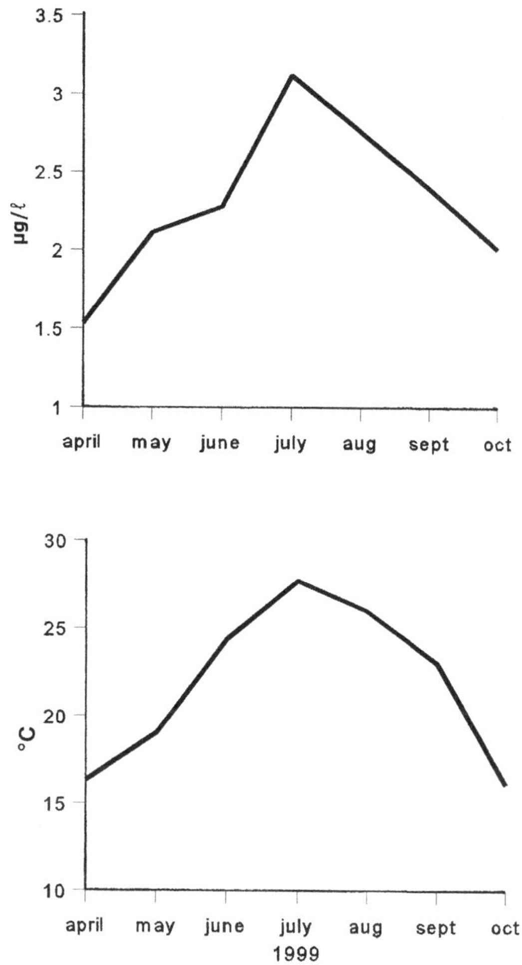

Fig. 11.1 Monthly mean surface temperature and chlorophyll-a concentration, Island Pond.

It was also observed during the course of the season that the chlorophyll concentration was consistently higher, usually by a factor of two, in the west sector of the pond than in the south sector, nearest the pond mouth. Furthermore, rate of growth of the oysters in the west sector, with respect to volume increase, was significantly higher than for those in the south sector. This degree of variation in food concentration in different parts of the pond might well account for the observed variations in oyster growth rate in different parts of the pond from year to year.

Observations have shown that chlorophyll and phytoplankton cell concentrations may increase markedly with depth in the pond, reaching a maximum at a depth of about 5 m. The phytoplankton population at this depth appears to consist largely of very small flagellates. The extent to which this potential food source may, as a result of vertical mixing, be utilized by the oysters suspended near the surface is unknown.

12 Oyster Culture in Tropical Regions

At the present time, the combined production of Mexico, Central and South America and Africa is less than 2% of total world oyster production, despite the extensive coast lines of these regions. Even though the tropical climate in large sectors of both continents favor an extensive growing season, and even though these regions suffer to varying degrees from an insufficiency of protein foods, oyster culture appears to have made little progress in recent decades.

Certain of the native mangrove oysters, so-called because of their habit of attaching to the roots of mangrove trees, can, under proper circumstances, grow rapidly year-around in response to constant favorable water temperatures. Also, the fecundity of oysters in general, in combination with a climate that stimulates reproduction during much if not all of the year, would seemingly provide a constant supply of seed oysters upon which to base an industry. However, these positive factors are often overwhelmed by a series of problems that have impeded most attempts at oyster culture in tropical areas.

Some of the major biological problems associated with tropical oyster culture have been reviewed in the preceding chapter. Added to these are certain socio-economic problems reflecting the absence of any infrastructure for developing an industry. Because of their perishability, live oysters are not easily handled or distributed in warm climates, particularly where major markets are few and far between and where transportation facilities are limited. In rural areas and coastal villages, oysters may be regularly consumed, but because of perishability, logistics and quality, market demand may be insufficient to justify culture efforts. Since they are not a luxury food and therefore usually do not command a high market price where sold, oyster culture in the tropics has not attracted a great deal of outside investment interest. Finally, in many of the so-called developing countries, lack of sanitation facilities result in chronic pollution problems in many coastal areas that might otherwise be suitable for oyster culture[12.1].

In 1995, Mexico produced about 35 000 t (whole weight) of oysters, which ranked it as sixth in total world production. Oyster landings in Mexico peaked during 1988 and 1989 at about 52 000 t, then declined (Fig. 12.1), apparently as a result of reduced market demand[12.2]. More recently, production has risen again, exceeding 40 000 t in 1997.

The species of major importance are the eastern oyster (*C. virginica*), which occurs naturally along Mexico's east coast (Gulf of Mexico), and

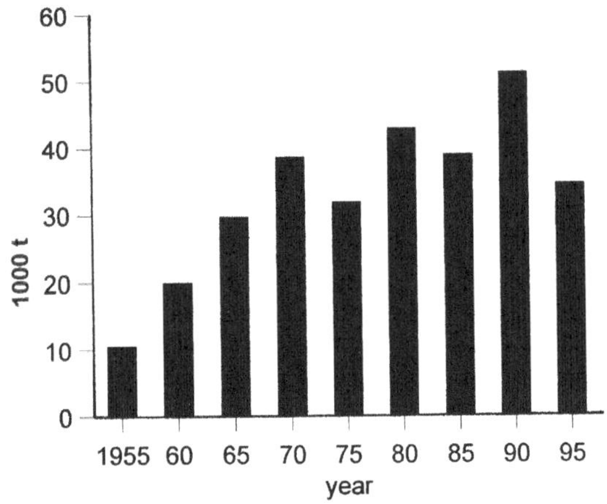

Fig. 12.1 Oyster production (whole weight) in Mexico.

the Pacific oyster (*C. gigas*), which was introduced to Mexico's west coast from oyster hatcheries in the United States. Two other species, the mangrove oyster (*C. corteziensis*) and rock oyster (*C. iridescens*), are both native to the Gulf of California and the west coast of Mexico[12.3]. Both these species are harvested commercially but in relatively low quantities. About 85% of total landings consist of eastern oysters from the Gulf of Mexico.

During the period 1955–1988, the growth in oyster production in Mexico was quite impressive, increasing from 10 000 t to over 50 000 t. However, production on the Pacific Coast as a result of new aquacultural techniques, including the establishment of shellfish hatcheries, has not fulfilled expectations, even though environmental conditions favorable to aquaculture and the potential for development exist[12.4]. There are presently four shellfish hatcheries operating in Mexico, with a combined output of about 42 500 000 spat per year[12.5], but hatchery production is not sufficient to eliminate the need for imported seed.

In Mexico, a series of fishing regulations issued in 1947 prohibited the harvest of oysters, as well as abalone, shrimp, lobster and several other species, by means other than through cooperatives[12.6]. At present, there are about 561 cooperatives engaged in oyster production[12.5], and these cooperatives can receive generous support from the state. However, members of an oyster cooperative may join another cooperative engaged in a different fishery on a seasonal basis, so that the continuity of a project may at times be disrupted.

Whatever the cause, the potential for the oyster industry in Mexico has not been realized. Many of the oyster resources of Mexico have been over-exploited, and pollution of the coastal lagoons by domestic and industrial wastes has been a major problem. In addition, oyster culture as opposed to oyster fishing is a relatively recent phenomenon in Mexico. The first attempts to improve growing areas by bottom conditioning and seed transplantation were not initiated until the late 1940s[12.7].

Virtually all oysters produced from the Gulf of Mexico are from the coastal lagoons, primarily in the State of Veracruz. These are large, shallow bodies of water, several in excess of 100 km in length but mostly less than 3 m in depth[12.2]. Water temperatures in these lagoons may range from 10°C or so in winter to 30°C or above, with a mean value around 20°C. For this reason, the oysters have an extended growing season and are generally in a constant state of gametogenesis. Most intensive spawning generally occurs during March–July and again in August–November. During the late summer and fall, heavy rainfall may reduce the salinity in the lagoons quite sharply, e.g. from 35‰ to 5‰, which often triggers spawning. However, such reductions in salinity are often accompanied by high oyster mortalities.

In the lagoons, culture generally takes the following form. In order to collect seed, concrete posts or wooden poles of otate (*Guadia amplexifolia*) are driven into the bottom and linked by horizontal traverses consisting of mangrove branches. The posts are positioned about 3.5 m apart, to form a rectangular module, and the horizontal beams are fastened about 2 m off the bottom. The cultch consists of oyster shells that are packed in plastic mesh bags or threaded through with nylon line to form strings. These are suspended from the horizontal beams. A major fouling organism is the barnacle (*Balanus* spp.), and it has been found that the strings are less prone to barnacle attachment if suspended in bundles. In some areas, oyster shell is simply spread over the bottom.

Within 2–3 months after setting, the seed have usually reached about 25 mm in size and are ready to be planted on firm bottom. The desired density is approximately 150 seed per m^2. Under normal conditions, mortality rate on the bottom is expected to be about 10% before harvesting. Major predators include the flatworm (*Stylochus* spp.), the blue crab (*Callinectes sapidus*) and carnivorous gastropods. Substantial losses may also be incurred after periods of heavy rain, which can result in low salinities and low oxygen concentrations because of the high organic content of the water. The oysters reach a marketable size of 70–75 mm within 6–12 months[12.2].

The oysters are harvested either by hand or by tongs. The great majority are shucked and sold raw, and the shells are replanted on the oyster beds. Because of uncertainty as to water quality, the great majority of the oysters are not exported but are consumed domestically.

On Mexico's Pacific coast and in the Gulf of California, two species are exploited intensively, the native mangrove oyster (*Crassostrea corteziensis*) and the rock oyster (*C. irridescens*). The mangrove oyster is a rapidly growing species that may reach a size of 75 mm during its first seven months[12.8]. It is generally found in sheltered mangrove areas and is tolerant of a wide range in salinity. Because this species has become increasingly scarce owing to intensive harvesting, efforts in hatchery production have been initiated in recent years. An oyster hatchery at San Blas, Nyarit, was established for this purpose, with expectations of an annual production of 60 000 000 spat.

In some areas, the mangrove oyster is cultured by suspending strings of shell from racks during the spawning season (August through October). After setting, the shells with seed attached are suspended from rafts, and the oysters are harvested nine months later[12.3].

Favorable year-around temperatures, strong tidal currents, and high primary productivity due to localized upwelling in many areas should be conducive to oyster production along Mexico's Pacific coast. Annual production from oyster rafts has averaged 5 t (whole weight) per raft, or 69 kg/m^2 of raft surface area[12.3].

In the State of Baja California, the Pacific oyster (*Crassostrea gigas*), obtained from hatcheries in Mexico and in the United States, has been of great importance to the industry. This species is capable of sustaining rapid growth in the relatively cold waters along the west coast of the Baja peninsula. At one of the larger oyster farms, mature oyster larvae are obtained from oyster hatcheries and placed in tanks in which plastic mesh bags of oyster shell are suspended. After setting has occured, the bags are held on intertidal racks for a period of 2 months, at which time the shells with seed attached are strung on wire and suspended from rafts. The oysters are generally harvested after 9 months, when 75–100 mm in size (Rene Islas Olivares pers. comm.).

A somewhat different method of culturing the Pacific oyster has been practiced in Baja California[12.9]. Single oysters about 25 mm in size, obtained monthly from a hatchery, are placed in plastic trays, 600 mm × 600 mm, which are stacked in tiers and suspended from a long-line. As they grow, the oysters are thinned periodically to a final density of about 80 per tray. Heavy biofouling and siltation requires constant maintenance. Silt is removed by hosing the trays with a pump, but exposure to air for a period of 2 days is required to eliminate many of the fouling organisms. Before marketing, in order to improve shelf life, the oysters are hardened by stacking the trays low in the intertidal zone for a period of 6–8 weeks.

The potential for culturing oysters and other marine species in Mexico, with over 7000 km of coast line and over 1 000 000 ha of estuaries and coastal lagoons along its Pacific Coast alone, is very impressive[12.4]. Other advantages include a year-around growing season and a ready domestic market for shellfish, and there are recent indications of increased production from the Gulf of Mexico, which accounts for nearly 90% of the nation's total. It is unfortunate that many of the coastal lagoons potentially favorable for shellfish culture are vulnerable to a variety of natural and anthropogenic hazards, including hurricanes and domestic and industrial wastes.

The species of oysters cultured in Central and South America and the Caribbean include the mangrove oyster (*Crassostrea rhizophorae*), the range of which extends from Mexico's Yucutan Peninsula southward throughout the Caribbean basin into northern South America, another so-called mangrove oyster (*C. brasiliana*), native to Brazil, which has been cultured experimentally near Sao Paulo, the Pacific oyster (*C. gigas*), which was introduced to Brazil and Venezuela from Japan and Korea, respectively, and the Chilean oyster (originally *Ostrea chilensis*, now designated *Tiostrea chilensis*), a flat oyster native to the west coast of South America. Only landings from Cuba, Dominican Republic, Brazil, Chile, and Venezuela are listed in the FAO 1995 Fishery Statistics, and it is assumed that commercial production in other countries in this region have been very small and therefore not reported.

The mangrove oyster (*C. rhizophorae*) resembles its northern relative, the eastern oyster, in general appearance and the northern limits of its range – the Yucatan Peninsula, the southern tip of Florida, and the Bahama Islands – mark the southern boundary of the range of *C. virginica*[12.10]. The mangrove oyster is so named because of its primary habitat, i.e. the roots of the red mangrove (*Rhizophorae mangle*). It can thrive in water of very high salinity, up to 38‰[12.11], and can tolerate even higher salinities for short periods. There is little seasonal variation in temperature throughout its range, and spawning may occur year-around. Oysters less than 4 months old may contain mature eggs and sperm, and, since spawning occurs throughout the year, the reproductive potential of this species is high.

Until recently, the leading oyster producer in this region has been Cuba, where, over the past few decades, production has averaged 2000–3000 t of mangrove oysters annually. In recent years, however, production has fallen below 2000 t. About half of Cuba's oysters have been produced through aquaculture, the remainder being harvested from wild populations[12.12]. The wild oyster fishery in Cuba is simply the process of collecting, by hand, oysters that have attached to mangrove roots and grown to marketable size[12.13].

The system of culture developed in Cuba is relatively primitive, utilizing inexpensive and readily available materials as much as possible. Wood obtained from mangroves is used to construct racks in relatively shallow water with good current flow. The posts are driven into the bottom, and horizontal beams connecting the posts are used to support spat collectors, usually consisting of bundles of mangrove branches. The bundles are arranged so that the terminal branches are exposed at low tide, which helps to reduce the amount of fouling by barnacles, tunicates, sponges and hydrozoans as well as predation by snails and crabs.

After the larval oysters settle on the branches and attach, the collectors may remain on the racks or are transferred to rafts, where growth is more rapid. In order to avoid excessive fouling, the bundles are removed from the water periodically and exposed to the air.

The growth of the juvenile oyster is rapid under normal prevailing temperatures, which, in Cuba, range from about 18°C in winter to 32°C in summer[12.11]. However, growth may be retarded by overcrowding, since setting occurs at varying intensity throughout the year. In Cuba, the minimum marketable size of 50 mm is usually reached within 3–5 months after settlement[12.11], while some individual oysters can reach 100 mm within 9 months. Similar rates of growth have been reported for the mangrove oyster in Puerto Rico[12.10] and Jamaica[12.14].

The harvesting of the oysters usually begins 5–6 months after the collectors were initially suspended from the racks. The larger oysters are picked off by hand, while the smaller ones are allowed to remain for additional growth. These are subsequently harvested at monthly intervals. After 8–9 months, the collectors are replaced with new ones. Once harvested, the oysters are shucked rather than being marketed for half-shell consumption, since their meats are apt to be small and often in poor condition.

Although this method of culture has been reasonably productive, it is not without problems. Because of the intermittent spawning throughout the year, the collectors often collect too many spat. Under these conditions, the oysters become stunted, and many small oysters attached to marketable oysters are wasted during harvesting. Other competitive fouling organisms include the flat tree oyster (*Isognomon alatus*), mussels (*Brachidontes* spp.) and barnacles (*Balanus* spp.). Many oysters are also lost to predatory crabs (*Callinectes bocourti* and *Panopeus lacustris*), gastropods (*Murex breviformis* and *Melongena melongena*) and fish, particularly the puffer, *Sphaeroides testudineus*)[12.11].

Attempts have been made to grow both the eastern oyster (*Crassostrea virginica*) and Pacific oyster (*C. gigas*) in Cuba to supplement the stocks of *C. rhizophorae*, but without success, presumably because of unfavorably high water temperatures[12.11]. In an effort to improve the quality of seed available to the growers, and improve the consistency in supply, a state oyster hatchery began operations a few years ago.

The largest mangrove oyster producer in the region currently is Venezuela, which harvested nearly 4000 t in 1995. Most of the production is from the harvest of wild stocks, but this species has been cultured in certain areas as well, using wooden rafts for suspension culture[12.15]. Spat collected on asbestos and cement sheets suspended beneath the surface reach marketable size within 6 months. However, the shells are very thin and brittle, and much care is required in the harvesting procedures.

The mangrove oyster is being cultured on a limited scale in Jamaica[12.14,12.16]. The major spawning period occurs from August to December, and care is taken to set out the cultch immediately after the settlement of barnacle larvae, which normally precedes the settlement of oysters. Although different types of cultch materials have been utilized, the least expensive and most readily available material used to date has been rubber squares cut from the walls of discarded automobile tires. These are strung on nylon monofilament line, separated from each other by sections of plastic tubing, and hung from racks constructed of bamboo and mangrove. In order to avoid fouling, the strings of cultch are exposed at low tide. After setting, the strings are transferred to rafts.

In Chile, the oyster industry is based upon two species, the native Chilean flat oyster (*Tiostrea chilensis*) and the Pacific oyster (*C. gigas*), introduced from the United States. The biology of the flat oyster is unusual in that the incubation period of the larvae may extend for 50 days or more within the mantle cavity of the female, while the duration of the swimming period is abbreviated to anywhere from a few days to a few hours. At time of setting, the size of the larva may be nearly twice that of advanced larvae of other species of oysters. This rather unusual larval period may in fact be a distinct advantage in maintaining populations in areas of swift tidal currents[12.17].

The range of this species extends from Ecuador to Chile, but the major area of production is the region of Chiloe in southern Chile. Spatfall is most intensive during the months of December, January and February[12.18]. Seed oysters are generally collected on cultch consisting of strings of mussel, scallop or clam shells, threaded through with nylon cord and separated by plastic tubing. These are arranged in rings and placed into trays holding adult oysters, which are suspended above the bottom. Additional shell strings are hung between the trays. Spawning begins when water temperatures reach about 12°C, and the mature larvae are expelled about 50 days later, when water temperatures have risen to about 21°C[12.19].

Because of the brevity of the larval swimming period, many of the larvae released in the trays attach to the shell collectors in, or in the immediate vicinity of, the same trays. When the juveniles reach a size of about 25 mm, the shell strings are transferred to rafts or long-lines. The oysters are marketable when they reach a size of about 50 mm, which may require 30–36 months depending upon water temperature.

During 1997, total oyster production in Chile amounted to about 3500 t. Nearly 90% of the harvest was accounted for by the Pacific oyster, a relatively recent introduction to Chile. This species, a faster grower than the flat oyster, is presently cultured successfully in Chile using the French system of plastic mesh bags supported off the bottom on racks.

In Africa, only Kenya, Senegal, and South Africa were listed in the 1995 catch statistics as oyster producers, and total production was only about 500 t. Oysters occur naturally, and are consumed, in other African countries such as Nigeria, but most of these apparently are collected and utilized for private consumption and seldom find their way to market. Species which have been cultured, albeit on an experimental basis, include *Crassostrea tulipia* (= *C. gasar*), the range of which extends from Senegal to Angola, *C. margaritacea*, which is native to South Africa, and the Pacific oyster, which was introduced to South Africa. Oyster production in the African countries is very limited at the present, and the majority are harvested from wild stocks.

Efforts to culture the native oyster (*C. tulipia*) in Sierra Leone[12.20] on the bottom were frustrated by heavy predation by boring snails. Experiments in off-bottom culture included the use of racks, from which strings of oyster shells, intended as cultch, were suspended, wooden trays with mesh bottoms, plastic mesh bags suspended from rafts, and rafts from which strings of oyster shells, with oysters attached, were suspended. The rafts were constructed of bamboo, floatation being provided by empty oil drums.

In Sierra Leone, spawning occurs with varying intensity year-around, the major peaks being in the spring and again in the fall. After setting, an average of 7–8 months of growth is required for the oysters to reach marketable size of about 75 mm.

A considerable range in salinity occurs in the culture areas of Sierra Leone because of the influence of the rainy season[12.20]. Salt concentrations may vary from 1‰ to 30‰ during the year, and the culture system has to be adjusted accordingly. The optimum schedule consists of collecting the spat in the early fall, culturing these throughout the winter when growing conditions are optimum, and harvesting the oysters in spring before the beginning of the rainy season. In this way, mortalities resulting from greatly reduced salinities during summer can be avoided. A 'seed-hardening' system has also been devised whereby spat collected during the rainy season are transferred to intertidal areas, thereby avoiding exposure to the minimum salinities coinciding with low tide.

13 Conclusions

In 1970, one of the leading authorities on the oyster industry in the United States wrote the following:

> 'By the year 2000, our industry probably will be growing large quantities of oysters in a mechanized closed system. Oysters will not be placed overboard in open waters at any time. Losses to predators and to smothering will be a thing of the past. Diseases of the young oysters will be known and controlled by drugs. Oysters will reach maturity in 12 months, with 3 or more months to fatten them for market. This rapid growth will be attained as a result of genetic studies to develop fast-growing strains, by controlling water temperatures at optimum levels, and force feeding with proper cultures of the right species of algae'[13.1].

Although these predictions did not seem unrealistic at the time, they clearly have not been realized by the year 2000. The concept of growing oysters in 'mechanized closed systems' on a commercial scale has been shown to be economically impractical beyond the oysters' juvenile stage, primarily because the volume of planktonic food that must be provided to sustain growth to market size is enormous, and hence costly, and is not justified by the market value of the product. Laboratory experiments in the culture of post-larval and juvenile oysters indicated that oysters in the 9–22 mm size range required about 80 000 000 algal cells as food per oyster per day in order to show an average weekly increase in shell height of 2 mm[13.2]. In order to sustain this weekly growth increment, it was found that the daily food requirement increased roughly in proportion to the oysters' shell surface area, and an oyster in the 40 mm size range would eventually require something in the order of 2 000 000 000 algal food cells each day. Later experiments in the culture of oysters in captivity indicated that about 450 g (dry weight) of algal cells would be required to produce 1.35–2.25 kg of live oysters (whole weight), at a cost ranging from about \$0.50 to \$2.00 per oyster for food alone[13.3]. Therefore oysters continue to be grown in the natural environment, and losses due to natural causes, such as predation, smothering, or adverse weather conditions, may be avoided to some extent but are nevertheless accepted as part of the process.

Rather than being reduced, the problem of disease has become increasingly widespread during the past 40 years, and much more remains to be known about these pathogens' biology and life cycles before the use of drugs can be considered. The primary strategy for coping with disease today is to produce, through selective breeding programs, oysters that are genetically tolerant if not resistant. However, it will require many generations of oysters, and therefore many years, for this approach to have a significant impact on the industry. Oysters bred for resistance to one particular parasite may have no defenses against attack by another.

Probably the most accurate of the predictions cited above was that relating to fast-growing strains. A small percentage of the oysters produced in Island Pond in recent years do in fact reach marketable size, or maturity, within 1 year, and many more are marketed at 18 months of age. This is partially due to the fact that growers are anxious to sell their product as soon as they possibly can in order to avoid disease-related losses, and relatively small oysters have become acceptable in the market. However, the growth rate of oysters in Island Pond has shown a significant improvement over the growth rate in earlier years, following selection of a different line obtained in 1985 as broodstock, and selecting the fastest growers for breeding stock each year may be a contributing factor. This improvement in growth rate has been achieved, however, without benefit of optimum temperature control or force feeding of selected species of algae once the oysters have completed metamorphosis.

The industry therefore has not, for the most part, developed along the lines forecast in 1970, and many of the traditional obstacles and natural hazards discussed above remain. An additional impediment, not only to oyster culture but to aquaculture in general in the United States, has been the difficulty in securing adequate control over natural bodies of water favorable for culture purposes. The coastal waters, and the fish and shellfish that dwell in them, have historically been regarded as a common resource that should always be open and available to public exploitation. Those who seek to assert exclusive rights over these waters, or parts thereof, threaten the so-called 'God-given' rights of both commercial and recreational fishing interests, and it is therefore not surprising that, in the majority of coastal communities, private efforts to engage in aquacultural enterprises are vigorously and more often than not successfully, opposed. This opposition is frequently joined by riparian landowners with homes near the water, who object to commercial activities on aesthetic grounds. Almost invariably, when the issue of whether or not to grant an aquaculture permit comes to a public vote, the permit seeker is greatly outnumbered and generally outvoted. Thus, according to a recent survey of aquaculturists in the north-east region of the United States, the greatest common constraint to aquaculture development was considered to be social and regulatory rather than technical or economic[13.4].

At least with respect to the shellfishing industries, there appears to be a marked difference in social attitudes between the United States and, for example, France or Japan. In both these countries, seafood is important, and oyster farming is strongly traditional, having become well established before the turn of the century – since the first half of the seventeenth century in Japan[13.5] – and having survived over the decades with relatively little change. The fact that the industries in both countries are not new but have been an important feature of their respective country's economy for many generations is significant, since they have not had to compete so vigorously with other more recent forms of coastal activity in order to establish a foot-hold and, over the years, have become politically more difficult to ignore. Both industries typically consist of numerous small family operations, or, in the case of Japan, cooperatives. Culture areas are leased from the government, and the leases are often passed on from one generation to the next.

As noted in an earlier chapter, Ocean Pond Corporation has been fortunate in encountering very little opposition to its culture operations in Island Pond. There have been occasional complaints about unsightly long-line buoys, which are camouflaged as much as possible by olive-green paint, and the odor of rotting algae in mid-summer, an occurrence that has nothing to do with culture activities, but the operation is generally accepted and even regarded with favor by some. However, in many coastal areas, there are just not that many sites favorable for raising shellfish that, if put to that purpose, would not conflict in some way with other interests.

Production in Island Pond is vulnerable chiefly to two natural phenomena, hurricanes and disease. Since changing the culture gear from surface trays to pearl nets suspended from long-lines, the chances of severe damage from violent storms are much reduced. However, the pond has not experienced a direct hit from a hurricane in recent years, and the chaos of entangled nets resulting from dragged moorings and broken lines would be considerable.

The threat of disease is more insidious. Each year, samples of oysters from the pond are sent to pathologists for examination. In 1995 and again in 1998, a small percentage of the oysters tested positively for *Perkinsus marinus* (Dermo), although there was no evidence of unusual mortality during either year. Since Island Pond is characterized by high summer water temperatures and a salinity range (normally 18–24‰) favorable to both *Perkinsus* and *Haplosporiodium*, the dense concentrations of oysters in the culture gear would seem to be particularly vulnerable.

Because the oysters are grown off-bottom in Island Pond, natural predation is not a serious factor. The flatworm, *Stylochus ellipticus*, is present in the pond and has been observed at times in the upwelling system and floating trays. This species has been known to destroy entire populations of newly-set oysters in salt ponds of Massachusetts. However, whatever losses it may cause in Island Pond are considered minimal. Perhaps more serious are the heavy infestations of *Polydora* that, when intensive enough, appear to weaken the shell and ultimately the attachment of the adductor muscle. However, such heavy infestations do not usually occur until the oysters are over 1 year old.

These features, combined with the absence of boating activities that might damage the culture gear, are important assets that have made Island Pond well suited for the culture of seed oysters.

References

Preface

P.1. Walford, L. A. (1958) *Living Resources of the Sea*. Ronald Press Company, New York.
P.2. Yonge, C. M. (1960) *Oysters*. Collins, London.
P.3. Iversen, E. S. (1968) *Farming the Edge of the Sea*. Fishing News (Books) Ltd, London.
P.4. Bardach, J. E., Ryther, J. H. & McLarney, W. O. (1972) *Aquaculture: the Farming and Husbandry of Freshwater and Marine Organisms*. Wiley-Interscience, New York.

Introduction

I.1. Loosanoff, V. L. (1956) On utilization of salt water ponds for shellfish culture. *Ecology*, **37**, 614–16.
I.2. Carriker, M. R. (1959) The role of physical and biological factors in the culture of *Crassostrea* and *Mercenaria*. *Ecological Monographs*, **29**, 219–66.

Chapter 1

1.1. Davis, H. C. & Calabrese, A. (1964) Combined effects of temperature and salinity on development and growth of larvae of *M. mercenaria* and *C. virginica*. *Fishery Bulletin*, **63**, 643–55.
1.2. Matthiessen, G. C. & Toner, R. C. (1966) *Possible Methods of Improving the Shellfish Industry of Martha's Vineyard, Dukes County, Massachusetts*. Marine Research Foundation, Edgartown, Massachusetts.
1.3. Matthiessen, G. C. (1960) Observations on the ecology of the soft clam, *Mya arenaria*, in a salt pond. *Limnology and Oceanography*, **5**, 291–300.
1.4. Korringa, P. (1976) *Farming the Flat Oysters of the Genus* Ostrea. Elsevier Scientific Publishing Company, Amsterdam.
1.5. Gaines, A. G. (1993) *Coastal Resources Planning and Management: Edgartown Great Pond, Edgartown, Massachusetts*. Marine Policy Center, Woods Hole Oceanographic Institution, Woods Hole, Massachusetts.

Chapter 2

2.1. Galtsoff, P. S. (1964) The American oyster *Crassostrea virginica* Gmelin. *Fishery Bulletin*, **64**, 1–480.
2.2. Stenzel, H. S. (1971) Oysters. In: *Treatise of Invertebrate Paleontology*, Part N, Bivalvia Vol. 3 (ed. R. C. Moore). University of Kansas and Geological Society of America, Inc., Boulder, Colorado.

2.3. Cai, Y. & Li, X. (1990) Oyster culture in the People's Republic of China. *World Aquaculture*, **21**, 67–72.

2.4. Chen, H. (1984) Recent innovations in the cultivation of edible mollusks in Taiwan. *Aquaculture*, **39**, 11–27.

2.5. Guo, X., Ford, S. E. & Zhang, F. (1999) Molluscan aquaculture in China. *Journal of Shellfish Research*, **18**, 19–32.

2.6. Bourne, N. (1979) Pacific oysters, *Crassostrea gigas* Thunberg, in British Columbia and the South Pacific islands. In: *Exotic Species in Mariculture* (ed. R. Mann), pp. 1–53. MIT Press, Cambridge, Massachusetts.

2.7. Sumner, C. (1980) Growth in Tasmanian oyster industry. *Australian Fisheries*, September, 11–15.

2.8. Rhodes, R. J. (1991) Chile's unique oyster. *Aquaculture Magazine*, March/April, 39–46.

2.9. Steele, N. (1964) *The Immigrant Oyster*. Warren's Quick Print, Olympia, Washington.

2.10. Maurin, C. & Le Dantec, J. (1979) The culture of *Crassostrea gigas* in France. In: *Exotic Species in Mariculture* (ed. R. Mann), pp. 106–22. MIT Press, Cambridge, Massachusetts.

2.11. Mann, R. (1979) Exotic species in mariculture: an overview of when, why, and how. In *Exotic Species in Mariculture* (ed. R. Mann), pp. 331–54. MIT Press, Cambridge, Massachusetts.

2.12. Imai, T. (1978) *Aquaculture in Shallow Seas; Progress in Shallow Sea Culture*. A. A. Balkema, Rotterdam.

2.13. Watters, K. W. & Prinslow, T. E. (1975) Culture of the mangrove oyster, *Crassostrea rhizophorae* Guilding, in Puerto Rico. *Proceedings of the World Mariculture Society*, **6**, 221–33.

2.14. Kamara, A. B., McNeil, K. & Quayle, D. B. (1979) Tropical mangrove oyster culture: problems and prospects. In: *Advances in Aquaculture* (eds T. V. R. Pillay & W. A. Dill), pp. 344–8. Fishing News (Books) Ltd., Farnham, England.

2.15. Kamara, A. B. (1982) Preliminary studies to culture mangrove oysters, *Crassostrea tulipa*, in Sierra Leone. *Aquaculture*, **27**, 285–94.

2.16. Bardach, J. E., Ryther, J. H. & McLarney, W. O. (1972) *Aquaculture: the Farming and Husbandry of Freshwater and Marine Organisms*, pp. 674–742. Wiley-Interscience, New York, NY.

2.17. Anant, S. (1982) Status of bivalve culture in Thailand. In: *Bivalve Culture in Asia and the Pacific* (eds F. B. Davy & M. Graham), pp. 73–8. International Development Research Centre, Ottawa, Canada.

2.18. Davy, F. B. & Graham, M. (1982) Workshop summary. In: *Bivalve Culture in Asia and the Pacific* (eds F. B. Davy & M. Graham), pp. 8–18. International Development Research Center, Ottawa, Canada.

2.19. Nell, J. A. (1993) Farming the Sydney rock oyster (*Saccostrea commercialis*) in Australia. *Reviews in Fisheries Science*, **1**, 97–120.

2.20. Curtin, L. (1971) Oyster farming in New Zealand. *New Zealand Marine Department, Fishery Research Report*, **7**, 1–99.

2.21. Yonge, C. M. (1960) *Oysters*. Collins, London.

2.22. Hidu, H. (1983) The new Maine oyster industry: status and problems. In: *Proceedings of the North American Oyster Workshop* (ed. K. K. Chew), pp. 233–48. World Mariculture Society, Special Publication No. 1. Louisiana State University, Baton Rouge.

2.23. Korringa, P. (1976) *Farming the Flat Oysters of the Genus* Ostrea, pp. 205–34. Elsevier Scientific Publishing Company, Amsterdam.

2.24. Sumner, C. (1972) Oysters and Tasmania, Part I. *Tasmanian Fisheries Research*, **6**, 1–15.

2.25. DiSalvo, L. H., Alarcon, E. & Martinez, E. (1983) Induced spat production from *Ostrea chilensis* (Philippi, 1845) in mid-winter. *Aquaculture*, **30**, 357–62.

2.26. Chanley, P. & Dinamani, P. (1980) Comparative descriptions of some oyster larvae from New Zealand and Chile, and a description of a new genus of oyster, *Tiostrea*. *New Zealand Journal of Marine and Freshwater Research*, **14**, 103–20.

2.27. Jeffs, A. G. & Creese, R. G. (1996) Overview and bibliography of research on the Chilean oyster *Tiostrea chilensis* (Philippi, 1845) from New Zealand. *Journal of Shellfish Research*, **15**, 305–11.

2.28. Breisch, L. L. & Kennedy, V. S. (1980) *A Selected Bibliography of Worldwide Oyster Literature*. Maryland Sea Grant Publication (UM-SG-80-11), University of Maryland, College Park, Maryland.

Chapter 3

3.1. Anon (1902–1990) *Fishery Statistics of the United States*, United States Department of Commerce, National Marine Fisheries Service, US Government Printing Office, Washington, DC.

3.2. Kochiss, J. M. (1973) *Oystering from New York to Boston*. Wesleyan University Press, Middletown, Connecticut.

3.3. Shaw, W. N. (1962) Raft culture of oysters in Massachusetts. *Fishery Bulletin*, **61**, 431–95.

3.4. Matthiessen, G. C. (1970) Production of seed oysters (*Crassostrea virginica*) in a brackishwater pond. In: *Coastal Aquaculture in the Indo-Pacific Region* (ed. T. V. R. Pillay), pp. 273–6. Fishing News (Books) Ltd., Farnham, England.

3.5. Loosanoff, V. L. (1966) Time and intensity of setting of the oyster, *Crassostrea virginica*, in Long Island Sound. *Biological Bulletin*, **130**, 211–27.

3.6. Matthiessen, G. C. & Toner, R. C. (1966) *Possible Methods of Improving the Shellfish Industry of Martha's Vineyard, Dukes County, Massachusetts*. Marine Research Foundation, Edgartown, Massachusetts.

3.7. Korringa, P. (1976) *Farming the Flat Oysters of the Genus* Crassostrea. pp. 187–204. Elsevier Scientific Publishing Company, Amsterdam.

3.8. Korringa, P. (1976) *Farming the Cupped Oysters of the Genus* Ostrea, pp. 107–18. Elsevier Scientific Publishing Company, Amsterdam.

3.9. Doumenge, F. (1990) Aquaculture in Japan. In: *Aquaculture, Volume 2*. (ed. G. Barnabe). Ellis Horwood, London.

Chapter 4

4.1. Dean, B. (1902) Japanese oyster-culture. *Bulletin of the United States Fisheries Commission*, **22**, 17–38.

4.2. Imai, T. (1978) *Aquaculture in Shallow Seas: Progress in Shallow Sea Culture*. A. A. Balkema, Rotterdam.

4.3. Mok, T. K. (1974) Observations on the growth of the oyster, *Crassostrea gigas* Thunberg, in Deep Bay, Hong Kong. *Hong Kong Fisheries Bulletin*, **4**, 45–53.

4.4. Stenzel, H. S. (1971) Oysters. In: *Treatise of Invertebrate Paleontology*, Part N, Bivalvia Vol. 3. (ed. R. C. Moore). University of Kansas and Geological Society of America, Inc., Boulder, Colorado.

4.5. Bardach, J. E., Ryther, J. H. & McLarney, W. O. (1972) Oyster culture. In:

Aquaculture: the Farming and Husbandry of Freshwater and Marine Organisms, pp. 674–742. Wiley-Interscience, New York.
4.6. Fujiya, M. (1970) Oyster farming in Japan. Helgoländer wissenschaftliche Meeresuntersuchungen, **20**, 464–79.
4.7. Koganezawa, A. (1979) The status of Pacific oyster culture in Japan. In: *Advances in Aquaculture* (eds T. V. R. Pillay & W. A. Dill), pp. 332–7. Fishing News (Books) Ltd., Farnham, England.
4.8. Fukuhara, O. (1981) Oyster culture in the Hiroshima Prefecture. *Tenth Joint Meeting, US/Japan Aquaculture Panel*, Delaware, US, 30–31 October 1981.
4.9. Kemp, S. & Wharton, R. (1995) Lessons from Japan. *Aquaculture Magazine*, **21**, 46–56.
4.10. Ventilla, R. F. (1984) Recent developments in the Japanese oyster culture industry. *Advances in Marine Biology*, **21**, 1–57.
4.11. Korringa, P. (1976) Farming the Pacific oyster (*Crassostrea gigas*) in Hiroshima Bay. In: *Farming the Cupped Oysters of the Genus* Crassostrea, pp. 153–82. Elsevier Scientific Publishing Company, Amsterdam.
4.12. Kim, K. H. (1962) Production of oysters in Korea. *Proceedings of the Indo-Pacific Fisheries Conference*, **10** (Sec. 2), 134–5.
4.13. Chen, H. C. (1984) Recent innovations in cultivation of edible mollusks in Taiwan. *Aquaculture*, **39**, 11–27.
4.14. Chew, K. K. (1994) Shellfish culture in Taiwan. *Aquaculture Magazine*, **20**, 69–75.
4.15. Nie, Zhong-Qing. (1982) Artificial culture of bivalves in China. In: *Bivalve Culture in Asia and the Pacific* (eds F. B. Davy & M. Graham), pp. 21–7. International Development Research Centre, Ottawa, Canada.
4.16. Cai, Y. & Li, X. (1990) Oyster culture in the People's Republic of China. *World Aquaculture*, **21**, 67–72.
4.17. Guo, X., Ford, S. E. & Zhang, F. (1999) Molluscan aquaculture in China. *Journal of Shellfish Research*, **18**, 19–32.
4.18. Qiu, Li-Qiang. (1982) Oyster culture in Guangdong. In: *Bivalve Culture in Asia and the Pacific* (eds F. B. Davy & M. Graham), pp. 27–8. International Development Research Centre, Ottawa, Canada.

Chapter 5

5.1. Castagna, M. & Kraeuter, J. N. (1981) Manual for growing the hard clam *Mercenaria*. *Special Report in Applied Science and Engineering*, Number 249. Virginia Institute of Marine Science, Gloucester Point, Virginia.
5.2. Haskin, H. H. & Ford, S. E. (1979) Development of resistance to *Minchinia nelsoni* (MSX) mortality in laboratory-reared and native stocks in Delaware Bay. *Marine Fisheries Review*, **41**, 54–63.
5.3. Bayes, J. C. (1981) Forced upwelling nurseries for oysters and clams using impounded water systems. In: *Nursery Culturing of Bivalve Mollusks* (eds C. Claus, N. DePauw & E. Jaspers), pp. 73–83. European Mariculture Society, Special Publication No. 7. Bredene, Belgium.

Chapter 6

6.1. Roughley, T. C. (1933) Life history of the Australian oyster, *Ostrea commercialis*. *Proceeding of the Linnaean Society of New South Wales*, **58**, 279–333.
6.2. Nell, J. A. (1993) Farming the Sydney rock oyster (*Saccostrea commercialis*) in Australia. *Reviews in Fisheries Science*, **1**, 97–120.

6.3. Hand, R. E., Nell, J. A., Smith, I. A. and Maguire, G. G. (1998) Studies on triploid oysters in Australia XI. Survival of diploid and triploid Sydney rock oysters (*Saccostrea commercialis*) (Iredale and Roughley) through outbreaks of winter mortality caused by *Mikrocytos roughleyi* infestation. *Journal of Shellfish Research*, **17**, 1129–35.
6.4. Meyers, J. A., Burreson, E. M., Barber, B. J. & Mann, R. (1991) Susceptibility of diploid and triploid Pacific oysters, *Crassostrea gigas* (Thunberg, 1793) and eastern oysters, *Crassostrea virginica* (Gmelin, 1791), to *Perkinsus marinus*. *Journal of Shellfish Research*, **10**, 433–8.
6.5. Wisely, B., Holliday, J. E. & Reid, B. L. (1979) Experimental deepwater culture of the Sydney rock oyster (*Saccostrea commercialis). Aquaculture*, **16**, 127–46.
6.6. Wisely, B., Holliday, J. E. & Reid, B. L. (1979) Experimental deepwater culture of the Sydney rock oyster (*Crassostrea commercialis). Aquaculture*, **17**, 25–32.
6.7. Hand, R. E., Nell, J. A. & Maguire, G. B. (1998) Studies on triploid oysters in Australia X. Growth and mortality of diploid and triploid Sydney rock oysters *Saccostrea commercialis* (Iredale and Roughley). *Journal of Shellfish Research*, **17**, 1115–27.
6.8. Sumner, C. (1972) Oysters and Tasmania, Part II. *Tasmania Fishery Research*, **8**, 1–12.
6.9. Thomson, J. M. (1959) The naturalization of the Pacific oyster in Tasmania. *Australian Journal of Marine and Freshwater Research*, **10**, 144–9.
6.10. Walker, T. (1993) Innovative oyster farm proves a hit with customers. *Austasia Aquaculture*, **7**, 7–11.
6.11. O'Sullivan, D. (1993) Oyster farming in Australia. *World Aquaculture*, **24**, 32–42.
6.12. Dinamani, P. (1971) Occurrence of the Japanerse oyster, *Crassostrea gigas* (Thunberg), in Northland, New Zealand. *New Zealand Journal of Marine and Freshwater Research*, **5**, 352–7.
6.13. Campbell, J. (1978) Aquaculture in New Zealand. *Fish Farming International*, **5**, 24–8.
6.14. Curtin, L. (1971) Oyster farming in New Zealand. *Fisheries Technical Report*, **72**, 1–99. New Zealand Marine Department, Wellington, New Zealand.
6.15. Young, A. & Serna, E. (1982) Philippines. In: *Bivalve Culture in Asia and the Pacific* (eds F. B. Davy & M. Graham), pp. 55–68. International Development Research Centre, Ottawa, Canada.
6.16. Angell, C. L. (1986) The biology and culture of tropical oysters. *ICLARM Studies and Reviews 13*. International Center for Living Aquatic Resource Management, Manila, The Philippines.
6.17. Anon. (1983) Oysters and mussels in the Philippines. *Fish Farming International*, **8**.
6.18. Brohmanonda, P., Mutarasint, K., Chongpeepien, T. & Amornjaruchit, S. (1988) Oyster culture in Thailand. In: *Bivalve Mollusk Culture Research in Thailand* (eds E. W. McCoy & T. Chongpeepien), pp. 31–9. Department of Fisheries, ICLARM Technical Report No. 19, Manila, The Philippines.
6.19. Ng, F. O. (1979) Experimental culture of the flat oyster *Ostrea folium* in Malaysian waters. *Malaysian Agriculture Journal*, **52**, 103–13.

Chapter 7

7.1. Raymont, J. E. G. (1947) A fish farming experiment in Scottish sea lochs. *Journal of Marine Research*, **6**, 219–27.
7.2. Cushing, D. H. (1975) *Marine Ecology and Fisheries*. Cambridge University Press, London.

7.3. Carpenter, E. J. & Dunham, S. (1985) Nitrogenous nutrient uptake, primary production, and species composition of phytoplankton in the Carmans River estuary, Long Island, New York. *Limnology and Oceanography*, **30**, 513–26.

Chapter 8

8.1. Korringa, P. (1976) *Farming the Flat Oysters of the Genus* Ostrea, pp. 187–204. Elsevier Scientific Publishing Company, Amsterdam.
8.2. Yonge, C. M. (1960) *Oysters.* Collins, London.
8.3. Balouet, G. & Poder, M. (1985) Current status of parasitic and neoplastic diseases of shellfish: a review. In: *Fish and Shellfish pathology* (ed. A. E. Ellis), pp. 371–80. Academic Press, London.
8.4. Goulletquer, P. & Heral, M. (1991) Aquaculture of *Crassostrea gigas* in France. In: *The Ecology of Crassostrea gigas in Australia, New Zealand, France, and Washington State* (eds J. Sutherland & R. Osman), pp. 13–19. Maryland Sea Grant College, College Park, Maryland.
8.5. Comps, S. (1988) Epizootic diseases of oysters associated with viral infections. In: *Disease Processes in Marine Bivalve Mollusks* (ed. W. S. Fisher), pp. 23–37. American Fisheries Society, Special Publication No. 18. Bethesda, Maryland.
8.6. Mann, R., Burreson, E. & Baker, P. (1991) The decline of the Virginia oyster industry in Chesapeake Bay: considerations for introduction of a non-endemic species, *Crassostrea gigas* (Thunberg, 1793). *Journal of Shellfish Research*, **10**, 379–88.
8.7. Maurin, C. & Le Dantec, J. (1979) The culture of *Crassostrea gigas* in France. In: *Exotic Species in Mariculture* (ed. R. Mann), pp. 102–22. MIT Press, Cambridge, Massachusetts.
8.8 Heral, M. (1990) Traditional oyster culture in France. In: *Aquaculture*, Volume 2 (ed. G. Barnabe), pp. 342–87. Ellis Horwood, London.
8.9. Goulletquer, P. & Heral, M. (1997) Marine mollusk production trends in France: from fisheries to aquaculture. In: *The History, Present Condition, and Future of the Molluscan Fisheries of North and Central America and Europe*, Volume III (eds C. L. MacKenzie, V. G. Burrell, A. Rosenfield & W. L. Hobart), pp. 137–64. US Department of Commerce, NOAA Technical Report, NMFS 127, Washington, DC.
8.10. His, E. & Robert, R. (1983) Developpement des veligers de *Crassostrea gigas* dans le bassin d'Arcachon. Etudes sur les mortalites larvaires. *Revue des Travaux de l'Institut des Pêches Maritimes*, **47**, 63–88.
8.11. Goulletquer, P., Wolowicz, M., Latala, A., Geairon, P., Huvet, A. & Boudry, P. (1999) Comparative analysis of oxygen consumption rates between cupped oyster spat of *Crassostrea gigas* of French, Japanese, Spanish and Taiwanese origins. *Aquatic Living Resources*, **12**, 233–302.
8.12. Cole, H. A. (1956) Benthos and the shellfish of commerce. In: *Sea Fisheries* (ed. M. Graham), pp. 139-206. Arnold Publishers Ltd, London.
8.13. Walne, P. R. (1974) *Culture of Bivalve Mollusks: 50 Years' Experience at Conway.* Fishing News (Books), West Byfleet, England.
8.14. Walne, P. R. & Helm, M. M. (1979) Introduction of *Crassostrea gigas* into the United Kingdom. In: *Exotic Species in Mariculture* (ed. R. Mann), pp. 83–105. MIT Press, Cambridge, Massachusetts.
8.15. Spencer, B. (1990) Farming the Pacific oyster in British waters. *Fish Farming International*, April 1990, 30–2.
8.16. Partridge, K. (1981) A manual for Irish oyster farmers. *Aquaculture Technical Bulletin*, **1**. National Board of Science and Technology, Dublin.

8.17. Wilson, J. H. & Simms, J. (1985) Gametogenesis and breeding of *Ostrea edulis* on the west coast of Ireland. *Aquaculture*, **46**, 307–21.
8.18. Mercer, J. P. (1981). Nursery culture of bivalve mollusks in Iceland – progress and problems. In: *Nursery Culture of Bivalve Mollusks* (eds C. Claus, N. DePauw & E. Jaspers), 189–95, European Mariculture Society, Special Publication No. 7. Bredene, Belgium.
8.19. Shatkin, G., Shumway, S. E. & Hawes, R. (1999) Considerations regarding the possible introduction of the Pacific oyster (*Crassostrea gigas*) to the Gulf of Maine: a review of global experience. *Journal of Shellfish Research*, **16**, 463–77.
8.20. Dijkema, R. (1988) Shellfish cultivation and fishery before and after a major flood barrier construction project in the south western Netherlands. *Journal of Shellfish Research*, **7**, 241–52.
8.21. Dijkema, R. (1997) Molluscan fisheries and culture in the Netherlands. In: *The History, Present Condition, and Future of the Molluscan Fisheries of North and Central America and Europe*, Vol. III (eds C. L. MacKenzie, V. G. Burrell, A. Rosenfield & W. L. Hobart), pp. 115–34. US Department of Commerce, NOAA Technical Report NMFS 127, Washington, DC.
8.22. Diaz, A. G. (1981) La ostricultura en Galicia. *Hoja del Mar*, **191**, 1–5.
8.23. Sanchez y Sanchez, M. (1956) The culture of oysters in northwestern Spain. *Journal du Conseil International pour l'Exploration de la Mer*, **22**, 197–9.
8.24. Cacares-Martinez, J. & Figueras, A. (1997) The mussel, oyster, clam and pectinid fisheries of Spain. In: *The History, Present Condition and Future of the Molluscan Fisheries of North and Central America and Europe*, Vol. III (eds C. L. MacKenzie, V. G. Burrell, A. Rosenfield & W. L. Hobart), pp. 165–90. US Department of Commerce, NOAA Technical Report NMFS 127, Washington, DC.

Chapter 9

9.1. Haskin, H. H. & Andrews, J. D. (1988) Uncertainties and speculations about the life cycle of the eastern oyster pathogen *Haplosporidium nelsoni* (MSX). In: *Disease Processes in Marine Bivalve Mollusks* (ed. W. S. Fisher), pp. 5–22. American Fisheries Society, Special Publication No. 18.
9.2. Leibovitz, L., Matthiessen, G. C. & Nelson, R. C. (1987) A preliminary study of diseases of cultured American oysters (*Crassostrea virginica*) during an annual growing cycle at the Cotuit Oyster Company. In: *Shellfish Diseases: Current Concerns in the Northeast* (ed. A. W. White). Woods Hole Oceanographic Institution, Technical Report WHOI-87-13.
9.3. Haskin, H. H. & Ford, S. E. (1979) Development of resistance to *Minchinia nelsoni* (MSX) mortality in laboratory-reared and native oyster stocks in Delaware Bay. *Marine Fisheries Review*, **41**, 54–63.
9.4. Ford, S. E. & Haskin, H. H. (1987) Infection and mortality patterns in strains of oysters *Crassostrea virginica* selected for resistance to the parasite *Haplosporidium nelsoni* (MSX). *Journal of Parasitology*, **73**, 368–76.
9.5. Matthiessen, G. C., Feng, S. Y. & Liebovitz, L. (1990) Patterns of MSX (*Haplosporidium nelsoni*) infection and subsequent mortality in resistant and susceptible strains of the eastern oyster, *Crassostrea virginica* (Gmelin, 1791) in New England. *Journal of Shellfish Research*, **9**, 359–65.
9.6. Ford, S. E. & Haskin, H. H. (1988) Management strategies for MSX (*Haplosporidium nelsoni*) disease in eastern oysters. In: *Disease Processes in Marine Bivalve Mollusks* (ed. W. S. Fisher), pp. 249–56. American Fisheries Society, Special Publication No. 18. Bethesda Maryland.

9.7. Chew, K. K. (1984) Recent advances in the cultivation of mollusks in the Pacific United States and Canada. *Aquaculture*, **39**, 69–81.
9.8. Stanley, J. G., Hidu H. & Allen, S. K. 1984. Growth of American oysters increased by polyploidy induced by the blocking of meiosis I but not meiosis II. *Aquaculture*, **37**, 147–55.
9.9. Matthiessen, G. C. & Davis, J. P. (1992) Observations on growth rate and resistance to MSX (*Haplosporidium nelsoni*) among diploid and triploid eastern oysters (*Crassostrea virginica* (Gmelin, 1791)) in New England. *Journal of Shellfish Research*, **11**, 449–50.
9.10. Barber, B. J. & Mann, R. (1991) Sterile triploid *Crassostrea virginica* (Gmelin, 1791) grow faster than diploids but are equally susceptible to *Perkinsus marinus*. *Journal of Shellfish Research*, **10**, 445–50.

Chapter 10

10.1. Smith, H. (1913) Oysters: the world's most valuable water crop. *National Geographic Magazine*, **24** (3), 257–81.
10.2. Ahmed, M. (1976) Speciation in living oysters. *Advances in Marine Biology*, **13**, 357–97.
10.3. Steele, E. N. (1964). *The Immigrant Oyster* (Ostrea gigas). Warren's Quick Print, Olympia, Washington.
10.4. Beattie, J. H., McMillin, D. & Weigardt, L. (1983) The Washington state oyster industry: a brief overview. In: *Proceedings of the North American Oyster Workshop* (ed. K. K. Chew), pp. 28–38. World Mariculture Society, Special Publication No. 1. Louisiana State University, Baton Rouge.
10.5. Loosanoff, V. L. (1955) The European oyster in American waters. *Science*, **121**, 119–21
10.6. Mackin, J. G. (1971) Oyster culture and disease. *Proceedings of the World Mariculture Society*, **1**, 35–8.
10.7. Kochiss, J. M. (1973) *Oystering from New York to Boston*. Wesleyan University Press, Middletown, Connecticut.
10.8. Elston, R. & Relyea, D. (1983) The oyster industry of Long Island and Long Island Sound region. In: *Proceedings of the North American Oyster Workshop* (ed. K. K. Chew), pp. 229–32. World Mariculture Society, Special Publication No. 1. Louisiana State University, Baton Rouge.
10.9. MacKenzie, C. L. (1983) To increase oyster production in the northeastern United States. *Marine Fisheries Review*, **45**, 1–22.
10.10. Loosanoff, V. L. & Davis, H. C. (1963) Rearing of bivalve larvae. *Advances in Marine Biology*, **1**, 1–136.
10.11. Davis, H. C. (1971) Design and development of an environmental controls system for culturing oyster larvae. In: *Artificial Propagation of Commercially Valuable Shellfish* (eds K. S. Price & D. Maurer), pp. 135–50. University of Delaware, Newark, Delaware.
10.12. Dugas, R. J., Pausina, R. V. & Voisin, M. (1983) The Louisiana oyster industry, 1980. In: *Proceedings of the North American Oyster Workshop* (ed. K. K. Chew), pp. 101–11. World Mariculture Society, Special Publication No. 1. Louisiana State University, Baton Rouge.
10.13. Shaw, N. (1962) Raft culture of oysters in Massachusetts. *Fishery Bulletin*, **65**, 431–95.
10.14. Aprill, G. & Maurer, D. (1976) The feasibility of oyster raft culture in coast estuaries. *Aquaculture*, **7**, 147–60.
10.15. Jeffries, L. R., Lutz, R. A. & Haskin, H. H. (1983) The Delaware Bay oyster industry: an overview. In: *Proceedings of the North American Oyster Workshop* (ed. K. K. Chew), pp. 220–8. World Mariculture Society, Publication No. 1. Louisiana State University, Baton Rouge.

10.16. Mann, R., Burreson, E. M. & Baker, P. (1991) The decline of the Virginia oyster fishery in Chesapeake Bay: considerations for introduction of a non-endemic species, *Crassostrea gigas* (Thunberg, 1793). *Journal of Shellfish Research*, **10**, 463–77.
10.17. Shatkin, G., Shumway, S. E. & Hawes, R. (1997) Considerations regarding the possible introduction of the Pacific oyster (*Crassostrea gigas*) to the Gulf of Maine: a review of global experience. *Journal of Shellfish Research*, **16**, 463–77.
10.18. Breese, W. P. (1979) Eyed larvae and the oyster industry. *Proceedings of the National Shellfish Association*, **69**, 202–3.
10.19. Burrell, V. C. (1985) Oyster culture. In: *Crustacean and Mollusk Aquaculture in the United States* (eds T. V. Heiner & E. E. Brown), pp. 235–73. AVI Publishing Company, Westport, Connecticut.
10.20. Perdue, J. (1980) A physiological approach to the summer mortality problem of Pacific oysters in Washington state. *Proceedings of the National Shellfish Association*, **70**, 133–4.
10.21. Welch, W. R. (1964) The European oyster, *Ostrea edulis*, in Maine. *Proceedings of the National Shellfish Assocication*, **54**, 7–24.
10.22. Glude, J. B. & Chew, K. K. (1982) Shellfish aquaculture in the Pacific northwest. *Alaska Sea Grant Report* 82–2, University of Alaska, Anchorage, Alaska.
10.23. Medcof, J. C. (1961) Oyster farming in the maritimes. *Bulletin of the Fisheries Research Board of Canada*, **131**, 1–158.
10.24. Chiasson, R. (1991) Malpeque oysters: trying to best a 100 year-old record. *World Aquaculture*, **22**, 36–40.
10.25. Jenkins, J. B. & MacKenzie, C. L. (1997) The molluscan fisheries of the Canadian Maritimes. In: *The History, Present Condition, and Future of the Molluscan Fisheries of North and Central America and Europe*, Vol. I. (eds C. L. MacKenzie, V. G. Burrell, A. Rosenfield & W. L. Hobart), pp. 15–44. US Department of Commerce, NOAA Technical Report NMFS 127, Washington, DC.
10.26. Gunn, R. C. (1983) A brief history of the oyster industry in Canada. In: *Proceedings of the North American Oyster Workshop* (ed. K. K. Chew), pp. 17–27. World Mariculture Society, Special Publication No. 1. Louisiana State University, Baton Rouge.
10.27. Bourne, N. F. (1997) Molluscan fisheries of British Columbia. In: *The History, Present Condition, and Future of the Molluscan Fisheries of North and Central America and Europe*, Vol. I. (eds C. L. MacKenzie, V. G. Burrell, A. Rosenfield & W. L. Hobart), pp. 115–30. US Department of Commerce, NOAA Technical Report NMFS 127, Washington, DC.

Chapter 11

11.1. Nixon, S. W. & Pilson, M. E. Q. (1983) Nitrogen in estuarine and coastal marine ecosystems. In: *Nitrogen in Estuarine and Coastal Marine Ecosystems* (eds E. J. Carpenter & D. G. Capone), pp. 565–648. Academic Press, New York.
11.2. Angell, C. L. (1986) *The Biology and Culture of Tropical Oysters*. ICLARM Studies and Reviews 13. International Center for Living Aquatic Resource Management, Manila, The Philippines.
11.3. Ryther, J. H. (1969) The potential of the estuary for shellfish production. *Proceedings of the National Shellfish Association*, **59**, 18–22.
11.4. Bardach, J. E., Ryther, J. H. & McLarney, W. O. (1972) *Aquaculture: the Farming and Husbandry of Freshwater and Marine Organisms*. Wiley-Interscience, New York.

Index

Note: page numbers in bold indicate figures.

Printed and bound by CPI Group (UK) Ltd, Croydon, CR0 4YY
16/09/2026
14963385-0003